INDEX OF SPECIAL SYMBOLS

Elementary Mathematical Analysis

Second Edition

Colin W. Clark

University of British Columbia

Wadsworth Publishers of Canada, Ltd.
10 Davis Drive
Belmont, California 94002

Mathematics Editor: John Kimmel
Production: Cobb/Dunlop Publisher Services, Inc.

Printed in the United States of America

   3 4 5 6 7 8 9 10   87 86 85 84

Library of Congress Cataloging in Publication Data

Clark, Colin Whitcomb, 1931-
  Elementary mathematical analysis, second edition

  Previous ed. published in 1972 as: The theoretical side of calculus.
  Includes bibliographical references and indexes.
  1. Mathematical analysis. I. Title.
QA300.C56 1982   515   81-4759
ISBN 0-534-98018-X   AACR2

Preface

The first edition of this book was titled *The Theoretical Side of Calculus*. This prosaic title indicated the objectives of the book, which were to present the logical ("epsilon-delta") basis of the theory of limits and their use in the calculus. This edition follows the first edition closely, with additional material on infinite series, additional exercises, and an application to the theory of differential equations.

The purpose of this text, then, is to provide a rigorous foundation for students familiar with the techniques of differentiation and integration, allowing them to proceed to more advanced topics such as multivariate analysis, real and complex analysis, differential equations, and so on.

The first three of the six chapters are especially elementary, covering the concepts of limit and continuity and the properties of the real number system. Limits are first discussed for the case of sequences (Chapter 1), and this material is applied to the theory of infinite series (Chapter 2). The transition to limits and continuity of functions (Chapter 3) is then fairly easy. Chapter 4 is devoted to proving the basic theorems about continuous functions. Calculus proper appears for the first time in Chapter 5, where the following topics are treated rigorously: differentiation and integration; uniform convergence of sequences and series of functions; transcendental functions. Chapter 6 gives a brief introduction to limits and continuity in *n* dimensions and concludes by proving the basic existence and uniqueness theorems for systems of ordinary differential equations—a proof that employs many of the results established throughout the text.

Of the two appendices, the second, on mathematical induction, is a prerequisite to the text proper, and should be studied by any student not familiar with this topic. Appendix I, on logic, is of course essential to any understanding of rigorous mathematical argument.

The material in this text is quite different from and needs to be studied differently from the material in most mathematics books. The emphasis here is entirely on the complete understanding of logical details, and not at all on the memorization of formulas and techniques. Many students will not previously have experienced the degree of mental concentration required to understand this material, and therefore the pace of instruction should at first be slow and methodical. Once some degree of confidence has been built up in handling "$\varepsilon - \delta$" and other logical arguments, the pace can be considerably quickened. To cover the entire text would take most of two semesters, but Chapters 1 to 4 and some of Chapter 5 could perhaps be done, with exceptional students, in a single semester.

I wish to thank the following colleagues for their useful comments: Arnold Grudin (Denison University), Brian Thompson (Simon Fraser University), Shaun Disney (The University of New South Wales), Bertram Yood (The Pennsylvania State University), David Leeming (University of Victoria), and Donald Schmidt (University of Northern Colorado). Finally, let me acknowledge the most valuable assistance of David Ryeburn (Simon Fraser University), who went over the manuscript of this edition with extreme care and contributed greatly to its improvement.

Contents

Historical Introduction

The Greek scholars were the first to consider seriously problems of continuity and infinity. These problems depend in turn on the basic concept of "number." Being excellent geometers, the Greeks attempted to include the concept of number in their geometry. Every line segment had a *length* (relative to a given unit length), which was expressed by a number; conversely every number was thought to be the length of some line segment. In this way, operations between numbers could be realized by means of geometrical "ruler and compass" constructions.

The construction of a line segment of *rational* length m/n (where m, n are integers, $n \neq 0$) is an easy matter. The hope that all numbers were rational was destroyed by the discovery, around 400 B.C., that $\sqrt{2}$ is irrational, even though it corresponds to a constructible line segment, the diagonal of a unit square. It is said that the Pythagorean mathematicians were so embarrased by the irrationality of $\sqrt{2}$ that they "classified" this information, and the first scholar to "leak" it to the public was poorly treated by his colleagues.

Three famous classical problems, those of "trisecting the angle," "doubling the cube," and "squaring the circle," were closely related to the question of realizing all numbers as lengths of constructible line segments. In modern terms these problems are equivalent to constructing, by ruler and compass, line segments of length $\cos 20°$ (for example), $\sqrt[3]{2}$, and π, respectively. In the nineteenth century all three constructions were shown to be impossible.† Thus

---

† P. L. Wantzel (1814–1848) proved that $\cos 20°$ and $\sqrt[3]{2}$ are nonconstructible. C. L. F. Lindemann (1852–1939) proved that π is "transcendental," meaning that π is not the root of any algebraic equation with integer coefficients, and hence is also nonconstructible.

the Greeks' concept of number was destroyed completely, and mathematicians were still faced with the task of defining number unambiguously and independent of geometrical considerations.

Toward the end of the nineteenth century several successful theories of the real number system were developed. The best known are those of Karl Weierstrass (1815–1897) and J. W. R. Dedekind (1831–1916). Both defined real numbers in terms of *infinite sets* of rational numbers. Later Georg Cantor (1845–1918) showed that the real numbers form an "uncountable" set, in the sense that, unlike the rational numbers, they cannot be placed in one-to-one correspondence with the set of positive integers. From this it follows that some (in fact most) real numbers cannot be obtained by any finite process from the rational numbers.

The work of Weierstrass and Dedekind showed that the real number system possesses a fundamental property not possessed by any smaller system (such as the rationals). This property, called *completeness*, can be described as follows: every nondecreasing, bounded sequence of real numbers "converges" to a real number. It follows from the completeness property that every decimal

$$m = a_0.a_1a_2a_3 \cdots ,$$

terminating or not, represents a real number, and conversely. This shows how to locate any real number, at least to any desired degree of accuracy, as a point on the "real line." Thus the Greek quest of identifying numbers with line segments was realized at last. Only the desire to obtain every number by (finite) construction turned out to be unrealizable.

Questions of continuity and infinity seem to have represented a complete mystery to the Greek scholars. The difficulties were clearly indicated by the famous paradoxes of Zeno (ca. 450 B.C.), of which we quote the following:

(a) The paradox of Achilles: In a race between Achilles and a tortoise, the latter has been given a head start. In order to catch up with the tortoise, Achilles must first cover half the distance separating them. Then he must cover half the remaining distance and so on. Hence, at any time, Achilles still has at least half the distance to go, and consequently is never able to pass the tortoise.

(b) The paradox of the arrow: Consider an arrow in flight. At any instant, the space occupied by the arrow is equal to the length of the arrow. Hence the arrow cannot undergo motion at any instant. Since time is composed of instants, no motion of the arrow is possible.

The paradox of Achilles seems to indicate that a continuous model of physical motion is self-contradictory, whereas the paradox of the arrow seems to show that a discrete model is equally self-contradictory. Speculation on matters of this sort later became a trademark of medieval theologism.

The modern era in mathematics and physics begins with one of the greatest geniuses of all time, Isaac Newton (1642–1727). In a period of less than two years, 1665–1666, there occurred the most cataclysmic intellectual revolution of all history. Working alone at his country home in order to escape the plague, Newton

(i) invented the differential and integral calculus, based on a study of infinite series;

(ii) established the basic laws of motion and of gravitation;

(iii) using the results of (i) and (ii), solved completely the ancient problem of the motion of the planets; and

(iv) discovered the nature of light.

Except for the work on light, Newton's discoveries went unpublished for over 20 years. Meanwhile G. Leibniz (1646–1716) discovered the calculus independently and published his theory. A stupid argument over priority arose between the followers of Newton and those of Leibniz. That Leibniz himself recognized Newton's accomplishment is obvious from his remark, "taking mathematics from the beginning of the world until the time of Newton, what he has done is much the better half."

In spite of the great success of their theories, neither Newton nor Leibniz was able to give a convincing explanation of the logical principles underlying the calculus. Both men struggled in vain with the concept of "infinitesimal"; Leibniz, in particular, committed several errors in questions relating to infinitesimals. The ensuing two centuries saw rapid and far-reaching developments in science and mathematics, and among practical-minded men the feeling grew that logical speculations were best avoided. In view of the tremendous success with which calculus could be applied to numerous physical phenomena, this viewpoint was clearly justified.

It is perhaps surprising, therefore, that any mathematicians remained sufficiently stubborn to insist on trying to understand the basic principles of their subject. But by the early nineteenth century certain developments in mathematical physics forced mathematicians to consider once more the basic problems of number, continuity, and infinity. During his work on the theory of heat, J. B. J. Fourier (1768–1830) emphasized the importance of *trigonometric series*,

$$b_0 + a_1 \sin x + b_1 \cos x + a_2 \sin 2x + b_2 \cos 2x + \cdots.$$

It soon became evident that such series behave quite differently from "power series,"

$$a_0 + a_1 x + a_2 x^2 + \cdots,$$

which were extensively used by Newton and his successors. We know now that if a power series *converges*, then the sum always represents a continuous (in fact, infinitely smooth) function. Fourier himself realized, on the other hand, that a trigonometric series could converge to a discontinuous function such as those shown below. Some mathematicians criticized Fourier's work by asserting

that such graphs do not represent actual "functions," that Fourier's concept of "convergence" was faulty, and so on. To settle these controversies, unambiguous definitions of such concepts as "function," "convergence," and "continuity" were needed. This brought mathematicians back to the problems of Zeno and to the necessity of developing calculus on a rigorous logical basis.

Among the earliest and most significant contributors to rigor in calculus was A. Cauchy (1789–1857). In his text *Leçons sur le calcul differentiel* (1829), Cauchy defined the derivative as the limit of a quotient:

$$\frac{dy}{dx} = \lim_{h \to 0} \frac{f(x+h) - f(x)}{h}.$$

Cauchy explained the meaning of limit in the following terms:

> When the successive values attributed to a variable approach indefinitely a fixed value so as to end by differing from it by as little as one wishes, this last is called the limit of the others.

Even this definition seems excessively vague by modern standards: the phrases "successive values," "variable," "approach indefinitely," "as little as one wishes," are all suggestive rather than precise.

Finally in 1872, H. E. Heine (1821–1881) presented the following formulation of the definition of the limit of a function $f(x)$ at x_0:

> "If, given any ε, there is an η_0 such that for $0 < \eta < \eta_0$ the difference $f(x_0 \pm \eta) - L$ is less in absolute value than ε, then L is the limit of $f(x)$ for $x = x_0$."

This statement, which is now the accepted definition of limit, is absolutely unambiguous. With minor modifications, it applies to many other kinds of limiting processes, including sequences and series of numbers and functions, functions of several variables, complex functions, and so on. The paradoxes of Zeno regarding time and motion disappear once the definition of continuity based on Heine's definition of limit is understood. There is probably no other instance in human intellectual history in which so much time and effort was spent merely to reach a satisfactory *definition*!

Given clear definitions of number, limit, continuity, and derivative, nineteenth-century mathematicians were able to provide a logically precise development of the calculus. The trigonometric series of Fourier became acceptable, thanks to studies by P. G. L. Dirichlet (1805–1859) and others. Of particular significance was the proof, by Weierstrass and B. Bolzano (1781–1848), that any continuous function $f(x)$ defined on a finite closed interval $a \leq x \leq b$ must assume finite extreme values.† It is clearly important to be able to deduce such "obvious" facts from given axioms and definitions.

In the twentieth century great developments in both pure and applied mathematics have been built on the foundations laid in the previous century. Although these developments are beyond the scope of this book, let us mention one particularly interesting example, the study of *spaces of infinite dimension*, initiated primarily by D. Hilbert (1862–1943). These spaces, whose properties obviously cannot be deduced by "geometrical" intuition, play an important role in modern physics, especially quantum mechanics.

By and large, the mathematics of this century is characterized by an un-limited capacity for generalization and abstraction. As in previous eras, much contemporary mathematical research seems to have little to do with practical affairs. Yet modern civilization is experiencing a rapidly increasing level of "mathematization," accelerated especially by the development of electronic computers. The major technical problems of the future are certain to require highly sophisticated modern mathematical techniques. People who clearly understand both the potentialities and the limitations of mathematical models will continue to be as valuable as they are rare.

(For further information on the history of calculus, see Carl B. Boyer, *A History of Mathematics*, Wiley (1968).)

---

† It should be pointed out that a certain amount of controversy is still attached to the meaning of theorems such as this, on the grounds that no *constructive* methods are possible for determining extreme points in general. See Appendix I.

1   *Sequences, Limits, and Real Numbers*

1.1   *Sequences*

The intuitive idea of an infinite sequence,

$$x_1, x_2, x_3, \ldots, x_n, \ldots, \tag{1.1}$$

can easily be given precise mathematical meaning.

Definition   *A sequence $\{x_n\}$ is a function defined for all positive integers $n = 1, 2, 3, \ldots$. Instead of the function notation $x(n)$, we use the subscript notation x_n for the nth **term** of the sequence.*

An example is the sequence $\{n/(n+1)\}$, in which $x_n = n/(n+1)$, or written in detail,

$$\frac{1}{2}, \frac{2}{3}, \frac{3}{4}, \ldots, \frac{n}{n+1}, \ldots.$$

Sequences are frequently specified by simply giving a formula for the nth term. For example if $x_n = (-1)^n$, the sequence $\{x_n\}$ is

$$-1, 1, -1, \ldots, (-1)^n, \ldots.$$

Another useful method for specifying a sequence is by means of a

1

recursion formula. As a simple example, consider the sequence

$$1, 3, 6, 10, ?, ?, \ldots.$$

(Such "sequences" used to occur on intelligence tests.) The terms given satisfy the following conditions:

$$x_1 = 1,$$

$$x_{n+1} = x_n + (n + 1) \quad (n \geq 1).$$

Any formula (such as the one just above) of the form

$$x_{n+1} = f_n(x_n) \tag{1.2}$$

is called a (two-term) recursion formula for the sequence $\{x_n\}$. It is sometimes possible to derive an explicit formula for x_n from such a recursion formula. Thus, in the example $x_{n+1} = x_n + (n + 1)$, we have:

$$x_1 = 1,$$
$$x_2 = 1 + 2,$$
$$x_3 = 1 + 2 + 3,$$
$$\cdot$$
$$\cdot$$
$$\cdot$$
$$x_n = 1 + 2 + 3 + \cdots + n$$
$$= \tfrac{1}{2}n(n + 1).$$

(On the last line we have used a well-known formula for $1 + 2 + 3 + \cdots + n$; see Appendix II.)

Exercises

1. Write out the first few terms of the sequences.

(a) $\{2^n\}$,

(b) $\{\cos n\pi\}$,

(c) $\left\{\dfrac{(-1)^n + 1}{2}\right\}$,

(d) $\{|n - 3| - 3\}$.

2.   Find simple sequences $\{x_n\}$ which begin as shown.

(a) $1, -\frac{1}{2}, \frac{1}{3}, -\frac{1}{4}, \ldots ,$ (b) $1, 3, 5, 7, \ldots ,$

(c) $1, 0, 1, 0, \ldots ,$ (d) $2, 5, 10, 17, \ldots ,$

(e) $4, 6, 10, 18, \ldots ,$ (f) $1, 2, 6, 24, \ldots .$

3.   Solve the following recursion formulas—that is, find x_n in general. Assume $x_1 = 1$.

(a) $x_{n+1} = x_n + 2,$ (b) $x_{n+1} = 1 - x_n,$

(c) $x_{n+1} = x_1 + x_2 + \cdots + x_n,$ (d) $x_{n+1} = \dfrac{n+1}{n} x_n,$

(e) $x_2 = -1$ and $x_{n+2} = x_n.$

4.   Given the information

$$x_1 = 1; \qquad x_{n+1} = x_n + n + 1 \quad (n \geq 1),$$

prove by mathematical induction that $x_n = \frac{1}{2}n(n+1)$ for all n.

5.   Use mathematical induction to prove that your answers to Exercise 3 are correct.

1.2   The Method of Iteration

Consider the problem of determining the value of $\sqrt{2}$ to a given degree of accuracy. If $x = \sqrt{2}$, then of course $x^2 = 2$. By simple algebra we can rewrite this in the strange form

$$x = \frac{x^2 + 2}{2x}. \tag{1.3}$$

Equation (1.3) is of the form

$$x = f(x). \tag{1.4}$$

The *method of iteration* for solving such an equation consists of constructing a sequence by the recursion formula

$$x_{n+1} = f(x_n). \tag{1.5}$$

Also x_1 has to be chosen to begin with. For Equation (1.3) we proceed as follows.

Let $x_1 = 1$ (a first "wild guess" for $\sqrt{2}$). Then using

$$x_{n+1} = \frac{x_n^2 + 2}{2x_n}, \tag{1.6}$$

we get

$$x_2 = \frac{1 + 2}{2} = 1.5,$$

$$x_3 = \frac{(1.5)^2 + 2}{2(1.5)} = 1.417,$$

$$x_4 = \frac{(1.417)^2 + 2}{2(1.417)} = 1.4142,$$

and so on.

It is a fact (which I hope you find amazing at this point) that the sequence $\{x_n\}$ constructed in this manner "converges" to $\sqrt{2}$ as n becomes large. It can even be shown that x_{n+1} is a better approximation to $\sqrt{2}$ than x_n is, by at least two decimal places. We will return to this question later. At the present time let us prove that

$$x_n \geq \sqrt{2} \quad \text{(for } n \geq 2\text{)}. \tag{1.7}$$

To see this, we use the inequality $a^2 + b^2 \geq 2ab$ (why is this inequality true?). Thus

$$x_{n+1} = \frac{x_n^2 + (\sqrt{2})^2}{2x_n} \geq \frac{2\sqrt{2}\,x_n}{2x_n} = \sqrt{2},$$

for $n \geq 1$, which proves (1.7).

The sequence $\{x_n\}$ of (1.6) will be studied in greater detail in Section 1.8.

Exercises

1. Find a recursion formula for calculating $\sqrt{a}$ $(a > 0)$. Try it on $\sqrt{3}$.

2. Let $\{x_n\}$ be the sequence of Equation (1.6). Using (1.7), show that $\{x_n\}$ is a nonincreasing sequence, in the sense $x_{n+1} \leq x_n$ (for $n \geq 2$).

3. Let $\{x_n\}$ be the above sequence. Prove that, actually, $x_n > \sqrt{2}$ $(n \geq 2)$. (Use mathematical induction, plus the fact that $a^2 + b^2 > 2ab$ unless $a = b$.)

4. Let $x = \sqrt{2}$; then $x = 2/x$. Investigate the usefulness of the recursion formula $x_{n+1} = 2/x_n$ for calculating $\sqrt{2}$.

*5.   Let $a > \sqrt{2}$. Define a' (Figure 1.1) to be the intersection with the x-axis of the line tangent to the curve $y = x^2 - 2$ at the point $(a, a^2 - 2)$. Find a'. What does this have to do with Formula (1.6)? (This construction is an example of "Newton's method," described in most calculus texts.)

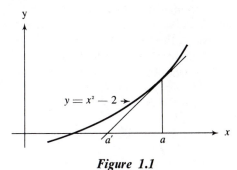

Figure 1.1

*6.   Use Newton's method to find a recursion formula for calculating $\sqrt[3]{2}$. Check your formula on a hand calculator.

1.3   *Properties of the Real Number System*

In the previous section we tacitly assumed that there *is* a number denoted by $\sqrt{2}$, or, in other words, that there is some (positive) number x such that $x^2 = 2$. Indeed such a "number" exists, but, as we now show, x cannot be a *rational number*, that is, not a "fraction" p/q with p and q integers. Consequently, the number system of ordinary mathematics must contain "irrational" numbers.

Theorem 1   *There is no rational number whose square equals 2.*

Proof   Suppose to the contrary that $x^2 = 2$ for some rational number $x = p/q$. Certainly we can assume that p and q have no common factors. The proof will be completed by "reductio ad absurdum." Namely, we will show that if $(p/q)^2 = 2$, then *both* p and q must necessarily be even integers, that is, p and q have a common factor of 2.

First note that $p^2 = 2q^2$. Hence p^2 is an even integer. Since the square of an odd integer is always odd (why?), this shows that p itself is even. Let $p = 2k$. Then we have $4k^2 = p^2 = 2q^2$, so that $q^2 = 2k^2$. But, as before, this implies that q must also be even. This completes the proof.   ∎

---

* Starred exercises are somewhat more difficult than average.

The existence of irrational numbers was one of the most confusing facts facing early mathematicians. Only in the nineteenth century was a successful theory developed, in which the entire real number system could be defined in terms of the *system of positive integers.*† Since this theory is quite complicated, it is not convenient to present it in an elementary book. Instead, we will treat the properties of the real number system as *axioms*, referring any interested reader to more advanced treatises which prove these "axioms" as theorems.‡ As far as calculus is concerned, the whole theory may be derived logically from these properties of the real number system.

Henceforth, we use the symbol $\mathbb{R}$ to denote the *set of all real numbers.* The notation $x \in \mathbb{R}$ means that x is a *member* of $\mathbb{R}$, that is, x is a real number. The symbol $\Rightarrow$ means "implies."

Properties of $\mathbb{R}$

1. Addition: $x, y \in \mathbb{R} \Rightarrow x + y \in \mathbb{R}$.

2. Multiplication: $x, y \in \mathbb{R} \Rightarrow xy \in \mathbb{R}$.

3. Commutative Laws: *For* $x, y \in \mathbb{R}$,
    (a) $x + y = y + x$,   (b) $xy = yx$.

4. Associative Laws: *For* $x, y, z \in \mathbb{R}$,
    (a) $x + (y + z) = (x + y) + z$,   (b) $x(yz) = (xy)z$.

5. Distributive Law: *For* $x, y, z \in \mathbb{R}$, $x(y + z) = xy + xz$.

6. Zero: *There is a unique number* $0 \in \mathbb{R}$ *such that*
    $$x + 0 = x \qquad \text{for every } x \in \mathbb{R}.$$

7. Subtraction: *Given* $x \in \mathbb{R}$, *there is a unique number*
    $-x \in \mathbb{R}$ *satisfying* $x + (-x) = 0$.

8. One: *There is a unique number* $1 \in \mathbb{R}$ *such that* $1 \neq 0$ *and*
    $$x \cdot 1 = x \quad \text{for every } x \in \mathbb{R}.$$

9. Division: *Given* $x \in \mathbb{R}$, $x \neq 0$, *there is a unique number*
    $x^{-1} \in \mathbb{R}$ *satisfying* $xx^{-1} = 1$.

---

† The famous remark "God made the integers; all the rest is the work of man," is attributed to L. Kronecker (1823–1891).

‡ See, for example, W. Rudin, *Principles of Mathematical Analysis*, 2d ed., McGraw-Hill (1964).

10. Order:

    (a) *Given* $x, y \in \mathbb{R}$, *then* $x = y$, $x < y$, *or* $y < x$; *no two can hold simultaneously.*

    (b) *If* $x < y$ *and* $y < z$, *then* $x < z$.

    (c) *If* $x < y$, *then* $x + z < y + z$ *for every* z.

    (d) *If* $x < y$, *then* $xz < yz$ *for every* $z > 0$.†

11. Completeness: (*See Section 1.8*).

We suppose that you are quite familiar with Properties 1 through 10, and these properties will be used without comment. Notice that Properties 1 through 10 cannot possibly describe $\mathbb{R}$ completely, because *the system* $\mathbb{Q}$ *of all rational numbers also satisfies Properties 1 through 10*. The same holds for certain other number systems (see Exercise 5 below). It is therefore interesting that a single additional property, called *completeness*, is sufficient to describe $\mathbb{R}$ precisely. This property is somewhat subtle, and we will discuss it in detail in Section 1.8.

If $x \in \mathbb{R}$, then $|x|$ (the *absolute value* of x) is defined by

$$|x| = \max(x, -x) = \begin{cases} x & \text{if } x \geq 0, \\ -x & \text{if } x < 0. \end{cases}$$

The following simple properties of the absolute value are easily checked, by considering the two cases $x \geq 0$ and $x < 0$:

$$x \leq |x|,$$
$$|x|^2 = x^2,$$
$$|x| = \sqrt{x^2},$$
$$|xy| = |x|\,|y|.$$

(1.8)

To conclude this section we derive a very important inequality, the "triangle inequality." It is an instructive, nontrivial exercise to fill in every logical detail of this proof, exhibiting every last application of Properties 1 through 10 (Property 11 not being required in the proof).

Theorem 2 (*The triangle inequality*)   *For any* $x, y \in \mathbb{R}$ *we have*

$$|x + y| \leq |x| + |y|.$$

(1.9)

---

† Of course $z > 0$ means the same as $0 < z$; such uses of inequality signs are assumed to be understood by the reader.

Proof First let us observe that

$$0 \le a < b \quad \text{implies} \quad a^2 < b^2.$$

This in turn shows that for $A, B \ge 0$

$$A^2 < B^2 \quad \text{implies} \quad A < B.$$

(Why?—this may require some thought!) Since both sides of the required inequality (1.9) are nonnegative, it is therefore sufficient to prove that

$$|x + y|^2 \le (|x| + |y|)^2.$$

But we have (using (1.8) several times):

$$
\begin{aligned}
|x + y|^2 &= (x + y)^2 \\
&= x^2 + 2xy + y^2 \\
&\le |x|^2 + 2|xy| + |y|^2 \\
&= |x|^2 + 2|x|\,|y| + |y|^2 \\
&= (|x| + |y|)^2. \quad \blacksquare
\end{aligned}
$$

Let us write the triangle inequality (1.9) in a different form. Since $a - b = (a - c) + (c - b)$, we see that

$$|a - b| \le |a - c| + |c - b|. \tag{1.10}$$

Definition *If $a, b \in \mathbb{R}$, we define the **distance between a and b** as*

$$\text{dist}\,(a, b) = |a - b|.$$

For example, dist $(3, 7) = |3 - 7| = 4$, whereas dist $(-3, 7) = |-3 - 7| = 10$. Inequality (1.10) therefore asserts that, for any point c, *the distance from a to b is not greater than the sum of the distances from a to c and from c to b.* This helps to explain why Inequality (1.9) is called the "triangle" inequality. The triangle inequality also holds for vectors (see Chapter 6), in which case actual triangles are involved.

Exercises

1. Show that $\sqrt[3]{2}$ is not rational.

2. Show that $\sqrt{3}$ is not rational.

3. Let $a \ne 0$ be a rational number and b an irrational number. Show that $a + b$ and ab are both irrational.

4. Which of the properties 1 through 10 of this section are possessed by:

    (a) $\mathbb{N}$, the set of positive integers?

    (b) $\mathbb{Z}$, the set of all integers?

    (c) $\mathbb{Q}^+$, the set of positive rational numbers?

5. Let $\mathbb{F}$ denote the set of all real numbers of the form

    $$a + b\sqrt{2},$$

    where a and b are rational numbers. Show that $\mathbb{F}$ has all the algebraic properties 1–10. (Properties 3, 4, 5, 6, 8, and 10 are obvious; concentrate on Properties 1, 2, 7, 9.)

6. Show that the following inequality is valid for all real numbers a, b:

    $$\big|\, |a| - |b| \,\big| \le |a - b|.$$

    (First substitute $x = a - b$, $y = b$ in the triangle inequality.)

7. When is it true that $|x + y| = |x| + |y|$?

8. Prove by induction: $\left|\sum\limits_{k=1}^{n} x_k\right| \le \sum\limits_{k=1}^{n} |x_k|.$

*9. Let $\mathbb{F}$ be the set of numbers given in Exercise 5. It is obvious that $\mathbb{F}$ is larger than the set $\mathbb{Q}$ of rational numbers, since $\sqrt{2} \in \mathbb{F}$ but $\sqrt{2} \notin \mathbb{Q}$. Show in turn that the set $\mathbb{R}$ is larger than $\mathbb{F}$. (Show that $\sqrt{3}$ is not in $\mathbb{F}$.)

1.4 The Limit of a Sequence (Introduction)

This section gives an intuitive introduction to the concept of the limit of a sequence. (Other types of limits are discussed in later chapters.) The intuitive idea is quite easy to grasp, and certain elementary calculations can be carried out without difficulty. On the other hand, the precise logical definition of limit, given in Section 1.6, is quite difficult. One cannot hope to proceed very far in mathematics, however, without mastering the concept of a limit in its exact formulation.

First then, for the intuitive idea: suppose $\{x_n\}$ is a given sequence. A certain real number a will be called the *limit* of the sequence $\{x_n\}$ if "x_n becomes arbitrarily close to a as n becomes large." If this is the case, we write

$$\lim_{n \to \infty} x_n = a,$$

or, alternatively,

$$x_n \to a \quad \text{as} \quad n \to \infty;$$

this last line is read: "x_n *approaches* (or *converges to*) *a as n approaches infinity*."

Example 1

Let $x_n = n/(n + 1)$. The terms of this sequence are

$$\frac{1}{2}, \frac{2}{3}, \frac{3}{4}, \dots,$$

and it is easy to guess that $\lim_{n \to \infty} x_n = 1$. This can be made even more transparent by writing

$$x_n = \frac{n}{n + 1} = 1 - \left(\frac{1}{n + 1}\right) \to 1 - 0 = 1 \quad \text{as} \quad n \to \infty.$$

Example 2

Let $x_n = (-1)^n$, that is,

$$-1, +1, -1, +1, \dots.$$

It should be clear that this sequence *has no limit*. (Some students may conjecture that the sequence has *two* limits, $+1$ and -1. But neither $a = +1$ nor $a = -1$ satisfies the requirement that "$(-1)^n$ becomes arbitrarily close to a as n becomes large." We will prove in Section 1.6 that a given sequence can have at most one limit.)

Example 3

Let $x_n = \sqrt[n]{n}$. A few terms of this sequence can be calculated approximately:

$$1.0, 1.41, 1.44, 1.41, 1.38, 1.34, \dots.$$

At this stage it is far from obvious whether $\lim_{n \to \infty} x_n$ exists, and if so, what its value is. We will return to this example also in Section 1.6.

The limits of a large class of sequences like that of Example 1 can be found by using a simple device:

Example 4

Let $x_n = (2n^2 - 5)/(n^2 + 4n)$. If we divide both numerator and denominator by the largest power of n appearing, we get

$$x_n = \frac{2 - \dfrac{5}{n^2}}{1 + \dfrac{4}{n}} \to \frac{2 - 0}{1 + 0} = 2 \quad \text{(as } n \to \infty\text{)},$$

or, in other words, $\lim_{n \to \infty} x_n = 2$. If this calculation is fairly obvious to you, then you already have a good intuitive idea of the limit concept.

Example 5

Let $x_n = n^2/2^n$. It is perhaps not at first obvious, but we can show that $\lim_{n \to \infty} x_n = 0$.

By the binomial theorem,

$$2^n = (1 + 1)^n$$

$$= 1 + n + \frac{n(n - 1)}{2} + \frac{n(n - 1)(n - 2)}{6} + \cdots + 1$$

$$> \frac{n(n - 1)(n - 2)}{6}.$$

Therefore, if $n > 2$,

$$\frac{n^2}{2^n} < \frac{6n^2}{n(n - 1)(n - 2)} = \frac{6n}{n^2 - 3n + 2},$$

which, by the method of Example 4, approaches zero as $n \to \infty$. Consequently $x_n \to 0$ as $n \to \infty$.

The question now arises, if an intuitive understanding of limits can be used to calculate answers to problems, why bother with a more formal logical approach? There are at least two responses to this question: first, one's intuition may not be strong enough to handle more difficult problems (such as $x_n = \sqrt[n]{n}$); second, two people's intuitions may give different answers to the same problem.

Example 6

Let $x_n = \left(\dfrac{n}{n+1}\right)^n.$

Since we know that $n/(n+1) \to 1$, we can guess that $x_n \to 1$ as $n \to \infty$. But this answer is incorrect, and in point of fact, it is known that $\lim_{n\to\infty} x_n = 1/e \simeq 0.368$. We put this example aside, together with Example 3, to return to it after a more careful study of limits. (See Section 1.9.)

Exercises

1. Decide which of the following sequences have limits, and find them by inspection when appropriate.

    (a) $\left\{\dfrac{n-2}{n}\right\},$

    (b) $\left\{\dfrac{n(n+4)}{4n(n+1)}\right\},$

    (c) $\left\{\dfrac{n^2+6}{n+6}\right\},$

    (d) $\left\{\sin\dfrac{n\pi}{2}\right\},$

    (e) $\left\{\dfrac{1}{n}\sin\dfrac{n\pi}{2}\right\},$

    (f) $\{2^{-n}\}.$

2. Calculate.

    (a) $\lim\limits_{n\to\infty}\dfrac{100-n}{100+n},$

    (b) $\lim\limits_{n\to\infty}\dfrac{2^n+n}{2^n-5},$

    (c) $\lim\limits_{n\to\infty}\dfrac{n^3}{2^n},$

    (d) $\lim\limits_{n\to\infty}\dfrac{2^n}{n^n},$

    (e) $\lim\limits_{n\to\infty}\dfrac{2^n+n^2}{3^n+n^2}.$

3. If $x_n = 0.99\ldots 9$ (n decimal places), what is $\lim_{n\to\infty} x_n$?

4. Concerning Problem 3, students sometimes ask the following question: "Supposing $x_n \to a$ as $n \to \infty$, what happens when $n = \infty$?" How does one answer this? Is it reasonable to say that a is the "last term of the sequence"?

5.   Let $x_n = \sqrt{n+1} - \sqrt{n}$. Show that $\lim_{n \to \infty} x_n = 0.$ $\left(Hint.\ \text{Show first that}\right.$

$$x_n = \frac{1}{\sqrt{n+1} + \sqrt{n}}\left.\vphantom{\frac{1}{1}}\cdot\right)$$

6.   Find

(a) $\lim_{n \to \infty} (\sqrt{n^2 + 2n + 1} - n)$,   (b) $\lim_{n \to \infty} (\sqrt{n^2 + n} - n)$.

(Neither limit is zero.)

1.5   The Limit of a Sequence (Numerical Examples)

Keep in mind that the formula $\lim_{n \to \infty} x_n = a$ is supposed to mean roughly that x_n becomes "very close" to a as n becomes "large." In this section we show how to answer the question: *How* large must n be in order to make the distance from x_n to a less than some given "tolerance." Remember that *the distance from x_n to a is equal to*

$$|x_n - a|.$$

Example 1

Let $x_n = (n-1)/(n+1)$, so that, obviously, $\lim_{n \to \infty} x_n = 1$. Given a "tolerance" of 0.01, find an integer N such that

$$|x_n - 1| < 0.01 \quad \text{for all } n \geq N.$$

Solution   By elementary algebra,

$$|x_n - 1| = \left| \frac{n-1}{n+1} - 1 \right|$$

$$= \left| \frac{-2}{n+1} \right|$$

$$= \frac{2}{n+1}.$$

Now $2/(n+1) < 0.01 = 1/100$ is valid if $n+1 > 200$. Let $N = 200$. We have shown that

$$|x_n - 1| < 0.01, \quad \text{provided } n \geq N.$$

Example 2

Clearly $\lim_{n \to \infty} 1/(n^3 - 3n + 1) = 0$. Given a "tolerance" of 10^{-3}, determine an integer N such that

$$\left| \frac{1}{n^3 - 3n + 1} \right| < 10^{-3}, \quad \text{provided } n \geq N.$$

Solution We simplify the denominator as follows:

$$n^3 - 3n + 1 > n^3 - 3n$$
$$\geq 9n - 3n \quad \text{if } n \geq 3 \quad (\text{why?})$$
$$= 6n.$$

Therefore,

$$\left| \frac{1}{n^3 - 3n + 1} \right| < \frac{1}{6n} \quad \text{if } n \geq 3.$$

Since $1/6n \leq 10^{-3}$ is valid if $n \geq 10^3/6$, which in turn is valid if $n \geq 200$, say, we can take $N = 200$.

Example 2 is worthy of careful study. Notice that we do not usually ask for the *smallest* possible value of N when dealing with limits. Any value of N that works for the necessary inequality is sufficient. (Of course if N works, so does any integer *larger* than N.) Keeping this in mind often allows much simplification in calculations. For example, one can show with some difficulty that $N = 12$ will work for Example 2 above; but this is completely unnecessary.

The following lemma is often useful.

Lemma   *Let $b = constant$. Then*

(i) $$n + b \leq 2n \quad \text{if } n \geq b,$$

(ii) $$n - b \geq \frac{n}{2} \quad \text{if } n \geq 2b.$$

The proof is completely trivial.

Example 3

Take a "tolerance" of $1/100$ and find N, as before, for the case

$$\lim_{n \to \infty} \left(\frac{n}{n^2 - 7} + \frac{5n + 1}{n + 5} \right) = 5.$$

Solution   To simplify the algebra we treat each fraction separately. Observing that $n/(n^2 - 7) \to 0$, we obtain

$$\left| \frac{n}{n^2 - 7} \right| \le \frac{n}{n^2/2} \qquad \text{if } n^2 \ge 14 \text{ or if } n \ge 4,$$

$$= \frac{2}{n} < \frac{1}{200} \quad \text{provided } n \ge 401 = N_1.$$

Next

$$\left| \frac{5n + 1}{n + 5} - 5 \right| = \frac{24}{n + 5}$$

$$< \frac{24}{n} \le \frac{1}{200} \quad \text{provided } n \ge 4800 = N_2.$$

Let us now choose $N = \max(N_1, N_2) = 4800$. By the triangle inequality we conclude that for $n \ge N$,

$$\left| \frac{n}{n^2 - 7} + \frac{5n + 1}{n + 5} - 5 \right| \le \left| \frac{n}{n^2 - 7} \right| + \left| \frac{5n + 1}{n + 5} - 5 \right|$$

$$< \frac{1}{200} + \frac{1}{200} = \frac{1}{100}.$$

Example 4

Observe that $\lim_{n \to \infty} 2n/(n + 3) = 2$. Let ε be an *arbitrary* positive "tolerance." Determine an integer N (depending on ε) such that

$$\left| \frac{2n}{n + 3} - 2 \right| < \varepsilon \quad \text{whenever } n \ge N.$$

Solution We proceed as in the previous examples:

$$\left| \frac{2n}{n+3} - 2 \right| = \frac{6}{n+3}$$

$$< \frac{6}{n} < \varepsilon \quad \text{provided } n > \frac{6}{\varepsilon}.$$

Hence, if we *let N be any integer* $> 6/\varepsilon$, we are through. (Notice that the smaller ε is, the larger N must be. This is typical of limits of sequences.)

Exercises

1. For each of the following sequences $\{x_n\}$, find the limit a by inspection. Then determine some integer N such that

$$|x_n - a| < \frac{1}{1000} \quad \text{for } every \ n \geq N.$$

   (a) $x_n = \dfrac{(-1)^n}{3n^2}$ (*Note*: $|(-1)^n| = 1$.),

   (b) $x_n = \dfrac{n+6}{n^2-6}$, (c) $x_n = \dfrac{n^2 + 10^{-3}}{n^2 - 10^{-3}}$,

   (d) $x_n = \dfrac{3n^3 - n + 8}{n(n-1)(n-2)}$,

   (e) $x_n = (0.9)^n$ (first find some k for which $x_k < 0.1$).

   (f) $x_n = \dfrac{n^3}{3^n}$, (g) $x_n = \sqrt{n^2 + 1} - n$,

   (h) $x_n = \dfrac{10^4}{n^{0.1} - 10^6}$.

2. Let $\varepsilon > 0$ be an arbitrary (unspecified) tolerance. For each example (a)–(d) of Exercise 1, determine an integer N, depending on ε, such that

$$|x_n - a| < \varepsilon \quad \text{for every } n \geq N.$$

3. Use the known fact that $\sin x < x$ for $x > 0$ to find an integer N such that

$$\left| \cos \frac{\pi}{n} - 1 \right| < \frac{1}{1000} \quad \text{for all } n \geq N.$$

1.6 The Limit of a Sequence

We come now to the heart of the subject matter of Chapter 1. First we will give the exact definition of the limit of a sequence. Then we will check, by means of simple examples, that the definition agrees with our intuitive understanding of limits. In the next section we will derive as theorems some important consequences of this definition.

Students of elementary mathematics frequently wonder why mathematicians are so fussy about giving precise definitions. In elementary courses, the basic concepts can usually be understood intuitively, and all problems encountered in the course can be solved on the basis of this intuitive understanding. This is especially true of a first course in calculus. But in studying more advanced mathematics, most students eventually come to a stage where they can no longer really follow the course unless they take the time to understand completely the underlying ideas of the subject. In fact, many students never seem to get over this "hump," and consequently, in a state of confusion, abandon mathematics. I hope that readers of this book will be able to continue their mathematical education unhindered by any confusion or misconception about the meaning of limits. (By the way, now would be a good time to look at Appendix I on Logic, especially the material in Section I.2.)

Suppose we know that $\lim_{n\to\infty} x_n = a$ for a certain given sequence $\{x_n\}$. Let $\varepsilon > 0$ be a given "tolerance." Then, as in the preceding section, it must be possible to determine an integer N such that $|x_n - a| < \varepsilon$ for all $n \geq N$. This idea is precisely what is taken for the *definition* of the limit.

Definition 1 *Let $\{x_n\}$ be a given sequence of real numbers, and let a be a given real number. Then*

$$\lim_{n\to\infty} x_n = a$$

means that:

for any given $\varepsilon > 0$, there is a corresponding integer N (which may depend on ε) such that

$$|x_n - a| < \varepsilon \quad \text{for every } n \geq N. \tag{1.11}$$

In understanding this definition, it is important to remember that $|x_n - a|$ represents the distance between x_n and a. The inequality $|x_n - a| < \varepsilon$ is equivalent to the double inequality (see Figure 1.2)

$$a - \varepsilon < x_n < a + \varepsilon. \tag{1.12}$$

The above definition can, therefore, also be worded as follows: $\lim_{n\to\infty} x_n = a$

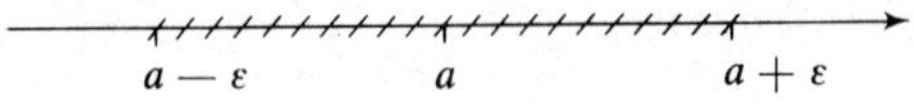

Figure 1.2

means that whatever interval† $(a - \varepsilon, a + \varepsilon)$ is taken (with $\varepsilon > 0$), *all the terms of the sequence $\{x_n\}$ except possibly the first $N - 1$ (depending on ε) lie within this interval.*

Example 1

$$\lim_{n \to \infty} \frac{1}{n} = 0.$$

To show that this is the case, we consider a given $\varepsilon > 0$. We must determine N (depending on ε) so that

$$\left| \frac{1}{n} - 0 \right| = \frac{1}{n} < \varepsilon \quad \text{for every } n \geq N.$$

Now $1/n < \varepsilon$, provided $n > 1/\varepsilon$. Let N be any integer larger than $1/\varepsilon$. Then, indeed,

$$\frac{1}{n} \leq \frac{1}{N} < \varepsilon \quad \text{if } n \geq N.$$

Example 2

$$\lim_{n \to \infty} \frac{1}{\sqrt{n}} = 0.$$

To see this, we note that $1/\sqrt{n} < \varepsilon$, provided $n \geq N > 1/\varepsilon^2$.

It is worth pausing to examine the form of the proof adopted in Examples 1 and 2. In order to prove, in any specific case, that $\lim_{n \to \infty} x_n = a$, the following steps must be taken:

(i) An *arbitrary* $\varepsilon > 0$ is considered;
(ii) An integer N is determined, by some form of mathematically logical reasoning, such that Condition (1.11) is satisfied.

---

† By the interval (h, k) we mean the set of all numbers lying between h and k exclusively, that is, the set of all $x \in \mathbb{R}$ satisfying $h < x < k$.

The degree of difficulty in any particular example will depend on how complicated the determination of N is. The next example is considerably more difficult than the previous two.

Example 3

$$\lim_{n \to \infty} \sqrt[n]{n} = 1.$$

As before, let $\varepsilon > 0$ be given. Let us write $y_n = \sqrt[n]{n} - 1$; then we must show that $|y_n| < \varepsilon$ for $n \geq N$.

By the binomial theorem (Appendix II), we have, since $y_n > 0$,

$$n = (1 + y_n)^n$$

$$= 1 + ny_n + \frac{n(n-1)}{2} y_n^2 + \cdots + y_n^n$$

$$> \frac{n(n-1)}{2} y_n^2 \quad \text{(if } n \geq 2).$$

Therefore for $n \geq 2$ we have

$$0 \leq y_n < \sqrt{\frac{2}{n-1}}.$$

But $\sqrt{2/(n-1)} \leq \sqrt{4/n}$ if $n \geq 2$, and this will be less than ε if $n > 4/\varepsilon^2$. Hence, if N is any integer greater than $4/\varepsilon^2$ (and $N \geq 2$), we have, as desired,

$$|y_n| < \varepsilon \quad \text{for all } n \geq N.$$

We will now prove that the *limit of a sequence is unique*, if it exists at all. You may object that this is obvious because if $\lim_{n \to \infty} x_n = a$ and $\lim_{n \to \infty} x_n = b$, then a and b are equal to the same quantity, and consequently $a = b$. We challenge you to find the flaw in this argument. It may not be obvious; see Exercise 5.

Theorem   *If a given sequence $\{x_n\}$ converges, then its limit is unique. (In other words, a sequence cannot have more than one limit.)*

Proof Suppose $\{x_n\}$ converges to both a and b. Then, given any $\varepsilon > 0$, there must exist an integer N_1 such that

$$|x_n - a| < \varepsilon \quad \text{for every } n \geq N_1$$

and an integer N_2 such that

$$|x_n - b| < \varepsilon \quad \text{for every } n \geq N_2.$$

Then, by the triangle inequality, we have

$$|a - b| = |a - x_n + x_n - b|$$
$$\leq |a - x_n| + |x_n - b| < 2\varepsilon, \quad \text{if } n \geq \max (N_1, N_2).$$

Therefore, given any $\varepsilon > 0$, we can show, by choosing a suitable value of n, that $|a - b| < 2\varepsilon$. But this is possible only if $|a - b| = 0$, that is, $a = b$. ∎

Definition 2   *A sequence that does not converge is said to **diverge**.*

Example 4

The sequence $\{(-1)^n\}$ diverges.

Suppose, on the contrary, we had $\lim_{n \to \infty} (-1)^n = a$. Then for every positive ε we would have

$$|(-1)^n - a| < \varepsilon$$

for all sufficiently large n. Taking a (large) *even* value of n, we conclude that

$$|1 - a| < \varepsilon.$$

But this inequality must be valid for every $\varepsilon > 0$, and therefore $a = 1$. On the other hand, the same reasoning using an odd value of n leads to the conclusion $a = -1$. This contradicts uniqueness of limits, and shows that $\{(-1)^n\}$ diverges.

The proofs just given for the theorem on uniqueness and for Example 4, although very simple, contain some subtle logical points. In both proofs, for example, we used the logical principle that if a statement S_n is true for *every* value of n in some range (such as $n \geq N$), then it is true for *any particular* value of n in this range.

We also used, in each proof, the mathematical principle that if $|x| < \varepsilon$ for every positive ε, then $x = 0$. This intuitively obvious fact can easily be derived from the axioms for $\mathbb{R}$; see Exercise 6.

Exercises

1. Show, by a proof similar to Example 1, that

$$\lim_{n \to \infty} \frac{1}{2n - 9} = 0.$$

2. Prove that

    (a) $\lim\limits_{n \to \infty} \dfrac{2n + 5}{n + 2} = 2,$ (b) $\lim\limits_{n \to \infty} \dfrac{(-1)^n}{n^2 + 3} = 0,$

    (c) $\lim\limits_{n \to \infty} \sqrt{\left(1 - \dfrac{1}{n}\right)} = 1,$

    (d) $\lim\limits_{n \to \infty} \dfrac{n^2}{2^n} = 0$ (cf. Example 5 of Section 1.4),

    (e) $\lim\limits_{n \to \infty} \dfrac{n^2 + 1}{2n^2 - 5} = \dfrac{1}{2},$ (f) $\lim\limits_{n \to \infty} \dfrac{2^n}{3^n} = 0,$

    (g) $\lim\limits_{n \to \infty} \dfrac{2^n + 3n}{3^n - 2n} = 0.$

3. Each of the following sequences either diverges or converges to zero. Decide which is which and give the complete ("ε-N") proof in each case.

    (a) $\{n\},$ (b) $\left\{\dfrac{(-1)^n + 1}{n^2}\right\},$

    (c) $\left\{\dfrac{1}{n} \sin n\right\},$ (d) $\left\{\sin \dfrac{n\pi}{3}\right\}.$

4. Does a constant sequence $(x_n = b$ for all $n)$ converge? Prove your answer.

5. Read the paragraph preceding the Theorem of this section, and criticize the argument suggested there for proving uniqueness of limits.

6. Prove that if $|x| < \varepsilon$ for every positive ε, then $x = 0$. (*Hint:* if $x \neq 0$, let $\varepsilon = \frac{1}{2}|x|$ and derive a contradiction.)

1.7 Elementary Theory of Limits

You will probably agree that the definition of limit given in the previous section is rather unwieldy. Consider, for example, the intuitively obvious formula

$$\lim\limits_{n \to \infty} \frac{n^2 + 6}{2n^2 + 1} = \frac{1}{2}; \tag{1.13}$$

it should not be necessary always to prove such formulas by "ε-N" arguments such as those given above. What is needed instead is a number of simple general principles which can be used in dealing with limits. The most important such principle is the following theorem, which can be summarized by saying that *all algebraic operations are preserved by the limit operation.*

Theorem 1   Let $\{x_n\}$ and $\{y_n\}$ be convergent sequences, with

$$\lim_{n \to \infty} x_n = a \quad \text{and} \quad \lim_{n \to \infty} y_n = b. \tag{1.14}$$

Then

(i) $\lim\limits_{n \to \infty} (x_n + y_n) = a + b.$

(ii) $\lim\limits_{n \to \infty} x_n y_n = ab.$

(iii) $\lim\limits_{n \to \infty} \dfrac{x_n}{y_n} = \dfrac{a}{b}$ if $b \neq 0$ and   no $y_n = 0.$

Before proving this theorem, let us show (in awesome detail) how it is used in practice on the example (1.13) given above:

$$\lim_{n \to \infty} \frac{n^2 + 6}{2n^2 + 1} = \lim_{n \to \infty} \frac{1 + \dfrac{6}{n^2}}{2 + \dfrac{1}{n^2}} \qquad \text{(by algebra)}$$

$$= \frac{\lim\limits_{n \to \infty}\left(1 + \dfrac{6}{n^2}\right)}{\lim\limits_{n \to \infty}\left(2 + \dfrac{1}{n^2}\right)} \qquad \text{(by (iii))}$$

(the limit in the denominator is not zero, as we shall see below). But now

$$\lim_{n \to \infty}\left(1 + \frac{6}{n^2}\right) = \lim_{n \to \infty} 1 + \lim_{n \to \infty} \frac{6}{n^2} \qquad \text{(by (i))}$$

$$= 1 + 6\left(\lim_{n \to \infty}\frac{1}{n}\right)^2 \qquad \begin{array}{l}\text{(by (ii) and}\\ \text{Exercise 4, Section 1.6)}\end{array}$$

$$= 1. \qquad \text{(by Example 1, Section 1.6)}$$

Similarly since $\lim_{n \to \infty} (2 + 1/n^2) = 2$, we get the answer (1.13) at last. Of course, this whole calculation is trivial and can be done in one's head; such may not be the case in more complicated examples, however.

Before giving the proof of Theorem 1 we introduce a simple but useful lemma. A sequence $\{x_n\}$ is said to be *bounded* if there is some constant M such that $|x_n| \leq M$ for all n. For example, $\{(-1)^n\}$ is a bounded sequence, but $\{n^2\}$ is not.

Lemma *Any convergent sequence is bounded.*

Proof Suppose that $\lim_{n \to \infty} x_n = a$. Taking $\varepsilon = 1$ in the definition of limit, we see that there exists a fixed integer N such that

$$|x_n - a| < 1 \quad \text{if } n \geq N.$$

By the triangle inequality

$$|x_n| \leq |x_n - a| + |a| < 1 + |a| \quad \text{if } n \geq N.$$

Define $M = \max(|x_1|, |x_2|, \ldots, |x_{N-1}|, 1 + |a|)$.† Then we have $|x_n| \leq M$ for every n. ∎

We now pass to the proof of part (ii) of Theorem 1, and ask you to try to prove parts (i) (which is a little easier) and (iii) (a little harder) yourself. Before writing out the formal "ε-N" proof, we try to see *why* (ii) must hold. From the hypothesis (1.14) we know that $x_n \to a$ and $y_n \to b$, or, in other words,

1. $|x_n - a|$ is small if n is large, and

2. $|y_n - b|$ is small if n is large.

We have to prove that $x_n y_n \to ab$; in other words,

3. $|x_n y_n - ab|$ is small if n is large.

To do this we proceed as follows:

$$
\begin{aligned}
|x_n y_n - ab| &= |x_n y_n - ay_n + ay_n - ab| \\
&= |(x_n - a)y_n + a(y_n - b)| \\
&\leq |x_n - a| \cdot |y_n| + |a| \cdot |y_n - b|.
\end{aligned}
\tag{1.15}
$$

Now we see that the terms on the last line are indeed small for large n. The formal proof follows.

---

† Note that a *finite* set of numbers certainly has a maximum element.

Proof of Theorem 1 (ii) First, by the Lemma $\{y_n\}$ is bounded; in other words, there is a positive number M such that

$$|y_n| \leq M \quad \text{for all } n.$$

Now let $\varepsilon > 0$ be given. By the definition of limit, there is an integer N_1 such that †

$$|x_n - a| < \frac{\varepsilon}{2M} \quad \text{for all } n \geq N_1. \tag{1.16}$$

Also, if $a \neq 0$, there is an integer N_2 such that

$$|y_n - b| < \frac{\varepsilon}{2|a|} \quad \text{for all } n \geq N_2. \tag{1.17}$$

Let $N = \max (N_1, N_2)$ if $a \neq 0$, and let $N = N_1$ if $a = 0$. Then by (1.15) we have, for all $n \geq N$,

$$|x_n y_n - ab| \leq |x_n - a| \cdot |y_n| + |a| \cdot |y_n - b|$$

$$< \frac{\varepsilon}{2M} \cdot M + |a| \cdot \frac{\varepsilon}{2|a|}$$

$$= \varepsilon$$

(the term $|a| \cdot \varepsilon/2|a|$ is omitted in case $a = 0$). This completes the proof of Theorem 1 (ii). ∎

Constructing "ε-N" proofs like this is rather difficult at first, but such proofs occur frequently in mathematical analysis.

Another important fact is that *order* ($\leq$) *is preserved by the limit operation.*

Theorem 2 *Let $\{x_n\}$ and $\{y_n\}$ be convergent sequences, and suppose $x_n \leq y_n$ for every n. Then*

$$\lim_{n \to \infty} x_n \leq \lim_{n \to \infty} y_n.$$

---

† It is important to understand the logic at this point. The symbol "ε," which has already been specified in this proof, does not now represent the same quantity as the "ε" used in (1.11) of the definition. Rather the "ε" of Equation (1.11) is now replaced by the quantity $\varepsilon/2M$ (which *is* a positive number). The same remark applies also to (1.17). Note also, however, that the integer N referred to by (1.11) depends on ε in general, and hence is not necessarily the same in (1.16) and (1.17). We have taken care of this possibility by using the notation N_1 and N_2 respectively.

Proof Let ε be a given positive number. Then there is an integer N such that

$$|x_n - a| < \varepsilon \quad \text{and} \quad |y_n - b| < \varepsilon \qquad \text{for } n \geq N,$$

where a and b are the limits of $\{x_n\}$ and $\{y_n\}$, respectively. This implies that

$$a - x_n < \varepsilon \quad \text{and} \quad y_n - b < \varepsilon \qquad \text{for } n \geq N$$

(explain!). Consequently, for $n \geq N$,

$$a - b = (a - x_n) + (y_n - b) + (x_n - y_n)$$
$$< 2\varepsilon,$$

since $x_n - y_n \leq 0$ by hypothesis. We have shown, therefore, that $a - b < 2\varepsilon$ for any positive ε. This means that $a - b \leq 0$, or

$$a = \lim_{n \to \infty} x_n \leq b = \lim_{n \to \infty} y_n. \quad \blacksquare$$

It should be noted that *strict* inequality $(x_n < y_n)$ is not necessarily preserved by the limit operation: see Exercise 10.

Next we calculate two especially important limits.

Theorem 3

(i) *If* $p > 0$, *then*

$$\lim_{n \to \infty} \frac{1}{n^p} = 0.$$

(ii) *If* $|a| < 1$, *then*

$$\lim_{n \to \infty} a^n = 0. \tag{1.18}$$

Proof The proof of (i) is easy, and is left to the reader. To prove (ii) we note that if $|a| < 1$ and $a \neq 0$, then $|a| = \dfrac{1}{1 + q}$ for some $q > 0$. Now, by the binomial theorem,

$$(1 + q)^n \geq 1 + nq > nq,$$

and therefore

$$|a^n - 0| = |a^n| = \frac{1}{(1 + q)^n} < \frac{1}{nq} < \varepsilon \quad \text{if } n > \frac{1}{\varepsilon q}.$$

This proves (ii) for $a \neq 0$; the case $a = 0$ is trivial. ∎

Finally, we calculate a useful limit of a different sort. The sum

$$1 + a + a^2 + \cdots + a^n = \frac{1 - a^{n+1}}{1 - a} \quad (a \neq 1) \tag{1.19}$$

is sometimes called a *geometric progression*. From Theorem 3 (ii) we obtain

$$\lim_{n \to \infty} (1 + a + a^2 + \cdots + a^n) = \frac{1}{1 - a} \quad (\text{if } |a| < 1). \tag{1.20}$$

Formula (1.20) is usually expressed in the notation of infinite series (see Chapter 2) as:

$$\sum_{n=0}^{\infty} a^n = \frac{1}{1 - a} \quad \text{if } |a| < 1.$$

Exercises

1.   This exercise is designed to provide you with some practice in the use of the logical phrases "for every (all, each)" and "for some" (or "there exists"). A more detailed treatment of these *logical quantifiers*, as they are called, is to be found in Appendix I.

The following 5 statements refer to a sequence $\{x_n\}$ and a real number a. The symbol N denotes a positive integer, and ε a positive number. For each statement (a)–(e), determine for which of the sequences (1)–(7) it is true.

Statements

(a) For every $\varepsilon > 0$ the inequality $|x_n - a| < \varepsilon$ holds for all but finitely many n.

(b) For every $\varepsilon > 0$ the inequality $|x_n - a| < \varepsilon$ holds for infinitely many n.

(c) For some N, the inequality $|x_n - a| < \varepsilon$ holds for all $n \geq N$ and for all $\varepsilon > 0$.

(d) There exists an $\varepsilon > 0$ such that $|x_n - a| \geq \varepsilon$ for all sufficiently large values of n.

(e) For every N there exists an $\varepsilon > 0$ such that $|x_n - a| < 1/N$ whenever $n \geq 1/\varepsilon$.

Sequences

(1) $x_n = n - |8 - n|$; $a = -8$.

(2) $x_n = 1 + \dfrac{(-1)^n}{n}$; $a = 1$.

(3) $x_n = (-1)^n + \dfrac{1}{n}$; $a = 1$.

(4) $x_n = 3$; $a = 3$.

(5) $x_n = \left(1 + \dfrac{1}{n}\right)^n$; $a = 2$.

(6) $x_n = n - |8 - n|$; $a = 8$.

(7) $x_n = 2^n/n^2$; $a = 1$.

2. Prove that if $\lim_{n\to\infty} x_n = a$, then $\lim_{n\to\infty} |x_n| = |a|$. (*Hint:* $\big|\,|x_n| - |a|\,\big| \leq |x_n - a|$ by the triangle inequality.)

3. Suppose $\lim_{n\to\infty} x_n = a$ and $x_n > 0$ for all n, and $a > 0$. Prove that $\lim_{n\to\infty} \sqrt{x_n} = \sqrt{a}$.

4. Prove that $\lim_{n\to\infty} cx_n = c \lim_{n\to\infty} x_n$. (Prove this directly, but note that it is a special case of Theorem 1(ii).)

5. Prove Theorem 1(i). (*Hint:* $|x_n + y_n - (a + b)| \leq |x_n - a| + |y_n - b|$.)

6. Prove the following Lemma: *If* $\lim_{n\to\infty} y_n = b \neq 0$, *then* $|y_n| \geq \tfrac{1}{2}|b|$ *for large n (i.e. for all $n \geq N_0$ for some integer N_0).*

7. Suppose that $\lim_{n\to\infty} y_n = b \neq 0$, and no $y_n = 0$. Use the lemma of Exercise 6 to show that $\lim_{n\to\infty} 1/y_n = 1/b$.

$$\left(\textit{Hint:} \left|\frac{1}{y_n} - \frac{1}{b}\right| = \frac{|y_n - b|}{|y_n|\,|b|}.\right)$$

8. Prove Theorem 1(iii). (*Hint:* Combine Exercise 7 with Theorem 1(ii).)

9. Prove by the method of Theorem 3(ii) that $\lim_{n\to\infty} na^n = 0$ if $|a| < 1$.

10. Let $\{x_n\}$ and $\{y_n\}$ be convergent sequences, and suppose $x_n < y_n$ for every n. Does it follow that $\lim_{n\to\infty} x_n < \lim_{n\to\infty} y_n$? Why?

11. Prove that if $\lim_{n\to\infty} x_n = a$, then $\lim_{n\to\infty} x_n^p = a^p$ for all positive integers p.

1.8   The Completeness Property

Given a sequence $\{x_n\}$, it is occasionally possible to show that it converges and to calculate the limit exactly by simple methods. But in many practical problems the reverse happens—namely, a sequential method is devised in order to calculate some otherwise unknown number. The question then arises whether the sequence obtained does converge, and if so, whether it converges to the desired number. (Also often of practical interest is the question: How *fast* does the sequence approach the desired number?)

Example 1

Let $\{x_n\}$ be the sequence of Section 1.2:

$$x_1 = 1, \qquad x_{n+1} = \frac{x_n^2 + 2}{2x_n}. \tag{1.21}$$

We can show now that *if this sequence converges*, then it must converge to $\sqrt{2}$. To see this, suppose

$$\lim_{n \to \infty} x_n = a.$$

Then also $\lim_{n \to \infty} x_{n+1} = a$, so that by (1.21)

$$a = \lim_{n \to \infty} x_{n+1} = \lim_{n \to \infty} \frac{x_n^2 + 2}{2x_n} = \frac{a^2 + 2}{2a}.$$

Solving for a, we find $a = \pm\sqrt{2}$. Since $x_n \geq \sqrt{2}$ for all n (see (1.7)), it must be the case that $a = \sqrt{2}$, as asserted.

To complete the discussion of the sequence (1.21) we only have to see why it must converge. First we recall from Section 1.2 the following facts:

(i) $x_n \geq \sqrt{2}$ for all $n > 1$,

(ii) $x_{n+1} \leq x_n$ for all $n > 1$.

Any sequence $\{x_n\}$ satisfying condition (ii) for all $n \geq 1$ is said to be *nonincreasing*. Thus our sequence $\{x_n\}$ is nonincreasing except for the first term. The sequence $\{x_n\}$ of (1.21) can be pictured as in Figure 1.3.

Figure 1.3

A basic property of the real number system $\mathbb{R}$, which completes our list given in Section 1.3, is:

11. Completeness Property of $\mathbb{R}$: *Every nonincreasing sequence of real numbers that is bounded below converges to some real number. Likewise, every nondecreasing sequence of real numbers that is bounded above converges to some real number.*

The completeness property is extremely important for the logical foundations of calculus (or "analysis"). Notice that the completeness property immediately applies to our sequence $\{x_n\}$ of (1.21), which must therefore converge to some real number. By what we have already proved, the limit must be $\sqrt{2}$:

$$\lim_{n \to \infty} x_n = \sqrt{2} \quad \text{if } x_n \text{ is given by (1.21).}$$

The completeness property of $\mathbb{R}$ thus guarantees the *existence* of a positive real number a such that $a^2 = 2$. (Recall that the lack of a *rational* number with this property caused concern to the Greek mathematicians.) Of course, as one would expect, completeness of $\mathbb{R}$ also guarantees the existence of nth roots $\sqrt[n]{x}$ $(x > 0)$ (see Section 4.5), as well as other important real numbers, such as π and e. Indeed, the completeness property of $\mathbb{R}$ is needed in order to *define* the standard elementary functions of calculus, such as $\sin x$, e^x, $\log x$, and so on (see Section 5.9).

Example 2

Consider the sequence

$$z_n = 1 + \frac{1}{2!} + \frac{1}{3!} + \cdots + \frac{1}{n!}.$$

Obviously z_n increases† with n. And since

$$\frac{1}{k!} = \frac{1}{2 \cdot 3 \cdot \, \cdots \, \cdot k}$$

$$\leq \frac{1}{2 \cdot 2 \cdot \, \cdots \, \cdot 2} = \frac{1}{2^{k-1}},$$

---

† The sequence $\{x_n\}$ is *increasing* (or *strictly increasing*) if $x_{n+1} > x_n$ for all n. Since an increasing sequence is certainly nondecreasing, the completeness property ensures that any increasing bounded sequence must converge.

we have

$$z_n \le 1 + \frac{1}{2} + \frac{1}{2^2} + \cdots + \frac{1}{2^{n-1}} = 2 - \frac{1}{2^{n-1}} < 2;$$

that is, the sequence $\{z_n\}$ is bounded above by 2. Therefore $\lim_{n \to \infty} z_n = a$ exists, as a consequence of the completeness property. In the next section we will show that $a = e - 1 \simeq 1.718$.

Another important example of an increasing, bounded sequence is

$$\left\{\left(1 + \frac{1}{n}\right)^n\right\};$$

this sequence is analyzed in detail in the next section.

The following theorem has already been used in several examples. Some will think that the result of the theorem is trivial; those who do may enjoy trying to find a proof which does not depend on the completeness property.

Theorem (Archimedean order property of $\mathbb{R}$)   *Let x be a real number. Then there is an integer n greater than x.*

Proof   Suppose, to the contrary, that every integer n is $\le x$. Then the sequence $\{x_n\}$ given by $x_n = n$ is increasing and bounded. Hence $\lim_{n \to \infty} x_n = a$ for some real number a. Then also $\lim_{n \to \infty} x_{n+1} = a$. But

$$x_{n+1} = 1 + x_n,$$

so that

$$\lim_{n \to \infty} x_{n+1} = 1 + a.$$

This implies that $a = 1 + a$ and, consequently, that $0 = 1$. Since $0 = 1$ is absurd, the proof is complete.   ▊

It follows from this theorem (and conversely; see Exercise 8) that:

If ε is a given positive number, then there exists an integer N such that $1/N < \varepsilon$.   (A)

If this property seems trivial to you, try to prove it using only Properties 1 through 10 of $\mathbb{R}$.

Next we present a brief discussion of decimals. By definition a decimal expansion is an expression of the form

$$m \cdot a_1 a_2 a_3 \ldots a_n \ldots,$$

where m is a nonnegative integer and each a_i is one of the digits from 0 to 9. Define a sequence $\{z_n\}$ by writing

$$z_n = m + \frac{a_1}{10} + \frac{a_2}{10^2} + \cdots + \frac{a_n}{10^n}.$$

Obviously $\{z_n\}$ is increasing; but also

$$z_n \le m + \frac{9}{10} + \frac{9}{10^2} + \cdots + \frac{9}{10^n} < m + 1,$$

so that $\{z_n\}$ is bounded. By the completeness property, $z = \lim_n z_n$ exists. Naturally we write

$$z = m \cdot a_1 a_2 a_3 \ldots a_n \ldots,$$

and we say that this is the *decimal expansion of z*.

Conversely it can be shown that every positive real number has a decimal expansion. For example, given z, we define m to be the greatest integer $\le z$. Then a_1 is defined to be the greatest integer $\le 10 (z - m)$, and so on. Except for the possibility of repeating 9's [e.g. $0.31999 \cdots = 0.32000 \cdots$], the decimal expansion is unique.

It is an interesting observation that the decimal expansion of any rational number must eventually repeat,† whereas the decimal expansion of any irrational number is nonrepeating. It may amuse you to try proving these assertions.

It would be difficult to overstress the importance of the completeness property in the study of analysis. This property will be used again and again in this book; in Chapter 4 we study the property itself with some care. I hope that you feel that the completeness property is intuitively reasonable. Actually if one begins with a clear definition of real number, it is possible to *prove* the completeness property.‡ However, since we have chosen to take $\mathbb{R}$ and its properties as a starting point, we accept the completeness property as an axiom.

---

† A decimal expansion is called *repeating* if the digits, from some point on, occur in a repeating group. For example, $2.7142142142 \cdots$ is a repeating decimal, which therefore is the decimal expansion of some rational number.

‡ cf. W. Rudin, *Principles of Mathematical Analysis*, 2d ed., McGraw-Hill (1964).

Exercises

1. If $\{x_n\}$ is defined by $x_1 = 1$ and

    $$x_{n+1} = \frac{x_n^2 + b}{2x_n} \quad (n \geq 1),$$

    where $b > 0$ is given, show in detail that $\lim_{n \to \infty} x_n = \sqrt{b}$.

2. Suppose the sequence $\{a_n\}$ satisfies

    $$a_1 = 1; \qquad a_{n+1} = \sqrt{1 + a_n} \quad (n \geq 1).$$

    (a) Prove by induction that $\{a_n\}$ is nondecreasing and that $a_n < 2$ for all n.
    (b) Show that $\lim_{n \to \infty} a_n$ exists, and find it.

3. Define a sequence $\{w_n\}$ by $w_1 = 1$ and

    $$w_{n+1} = \frac{2}{w_n^2} \quad \text{if } n \geq 1.$$

    Show first that *if* $\lim_{n \to \infty} w_n = w$ exists, then $w^3 = 2$. But then show that the sequence $\{w_n\}$ does *not* approach any limit. (This shows that one shouldn't jump to conclusions!)

4. (a) Prove by induction that

    $$1 + \frac{1}{2^2} + \frac{1}{3^2} + \cdots + \frac{1}{n^2} \leq 2 - \frac{1}{n} \quad (n = 1, 2, 3, \ldots).$$

    (b) What can you say about

    $$\lim_{n \to \infty} \left(1 + \frac{1}{2^2} + \frac{1}{3^2} + \cdots + \frac{1}{n^2}\right)?$$

5. Show by induction that

    $$1 + \frac{1}{2^2} + \frac{1}{3^2} + \cdots + \frac{1}{n^2} \leq \frac{7}{4} - \frac{1}{n} \quad \text{if } n \geq 2.$$

    Hence, answer Exercise 4(b) more precisely.

*6. Expanding on the ideas of Exercises 4 and 5, show how to calculate the limit 4(b) to within a given error.

7. Let $x_n = \dfrac{1 \cdot 3 \cdot 5 \cdot \cdots \cdot (2n - 1)}{2 \cdot 4 \cdot 6 \cdot \cdots \cdot 2n}$. Show that

$$\lim_{n \to \infty} x_n = x \quad \text{exists and} \quad 0 \leq x \leq \tfrac{1}{2}.$$

(In fact $x = 0$, but this is tricky to show; see Exercise 12 below.)

8. Show that the Archimedean order property implies Condition (A) and conversely.

9. Detect the use of (A) in proving that $\lim_{n \to \infty} 1/n = 0$ (see Example 1, Section 1.6).

10. Show that if $a = p/q$ is a rational number, then there exists an integer $N > a$; show this without using the completeness or Archimedean order properties of $\mathbb{R}$.

*11. Consider the sequence $\{x_n\}$ defined by

$$x_1 = 0, \qquad x_{n+1} = \frac{1 - x_n}{2}.$$

Calculate the first few terms; then discuss convergence of the sequence. (*Hint:* Consider the "alternate" subsequences $\{x_{2n-1}\}$ and $\{x_{2n}\}$.)

*12. (a) Let $y_n = \dfrac{2 \cdot 4 \cdot 6 \cdot \cdots \cdot 2n}{3 \cdot 5 \cdot 7 \cdot \cdots \cdot (2n + 1)}$. Show that $\lim\limits_{n \to \infty} y_n = y$ exists and $0 \leq y \leq \tfrac{2}{3}$.

(b) For x as given in Exercise 7, and y as in (a) above, show that

$$xy = \lim_{n \to \infty} x_n y_n = 0.$$

Since $x \leq y$ (why?) this proves that $x = 0$.

1.9 The Base of Natural Logarithms

Consider the sequence

$$x_n = \left(1 + \frac{1}{n}\right)^n. \tag{1.22}$$

We are going to prove that $\{x_n\}$ is a bounded, increasing sequence, which consequently must converge, according to the completeness property. Hence we can make the following statement.

Definition $\lim\limits_{n\to\infty}\left(1+\dfrac{1}{n}\right)^n = e.$ *The number* e *is called the* **base of natural logarithms.**

Calculus students may be familiar with e because of the important formulas

$$\frac{d}{dx}\,e^x = e^x; \qquad \frac{d}{dx}\,\log_e x = \frac{1}{x},$$

which are discussed in every calculus text.

We show first that $\{x_n\}$ is bounded. By the binomial theorem, we have

$$x_n = \left(1+\frac{1}{n}\right)^n = 1 + \frac{n}{1!}\cdot\frac{1}{n} + \frac{n(n-1)}{2!}\left(\frac{1}{n}\right)^2 + \cdots + \left(\frac{1}{n}\right)^n$$

$$< 1 + \frac{1}{1!} + \frac{1}{2!} + \cdots + \frac{1}{n!}$$

$$\le 1 + \left(1 + \frac{1}{2} + \frac{1}{2^2} + \cdots + \frac{1}{2^{n-1}}\right)$$

$$< 1 + 2 = 3 \tag{1.23}$$

(here we used the simple inequality $1/k! \le 1/2^{k-1}$). Thus $x_n < 3$ for every n; that is, $\{x_n\}$ is a bounded sequence. Next we show that $\{x_n\}$ is an increasing sequence. By the binomial theorem once more,

$$x_n = \left(1+\frac{1}{n}\right)^n$$

$$= 1 + 1 + \frac{n(n-1)}{2}\cdot\frac{1}{n^2} + \frac{n(n-1)(n-2)}{3!}\cdot\frac{1}{n^3} + \cdots + \frac{n!}{n!}\cdot\frac{1}{n^n}$$

$$= 2 + \frac{1}{2}\left(1-\frac{1}{n}\right) + \frac{1}{3!}\left(1-\frac{1}{n}\right)\left(1-\frac{2}{n}\right) + \cdots$$

$$+ \frac{1}{n!}\left(1-\frac{1}{n}\right)\cdots\cdots\left(1-\frac{n-1}{n}\right). \tag{1.24}$$

In the analogous formula for x_{n+1}, every term on the right-hand side (except the first term) increases in value and an additional term appears on the right. Therefore, $x_{n+1} > x_n$. We have now shown that the sequence $\{x_n\}$ is both bounded and increasing; consequently $e = \lim_{n \to \infty} x_n$ exists.

We now derive another formula for e, which will be useful for calculating e numerically. Consider the sequence $\{y_n\}$ defined by

$$y_n = 1 + \frac{1}{1!} + \frac{1}{2!} + \cdots + \frac{1}{n!}.$$

Obviously $\{y_n\}$ is an increasing sequence. It follows from (1.23) that $y_n < 3$ for all n. Therefore, by the completeness property,

$$f = \lim_{n \to \infty} y_n$$

exists. We now want to prove that

$$f = e.$$

First, from (1.23), we have

$$x_n = \left(1 + \frac{1}{n}\right)^n < 1 + \frac{1}{1!} + \frac{1}{2!} + \cdots + \frac{1}{n!} = y_n.$$

Therefore,

$$e = \lim_{n \to \infty} x_n \le \lim_{n \to \infty} y_n = f. \tag{1.25}$$

Conversely, let m be a given integer. Suppose $n \ge m$; taking the first m terms in (1.24) we get

$$x_n > 2 + \frac{1}{2!}\left(1 - \frac{1}{n}\right) + \frac{1}{3!}\left(1 - \frac{1}{n}\right)\left(1 - \frac{2}{n}\right) + \cdots$$

$$+ \frac{1}{m!}\left(1 - \frac{1}{n}\right) \cdots \left(1 - \frac{m-1}{n}\right).$$

Keeping m fixed, we let $n \to \infty$ on both sides of this inequality, obtaining

$$e \ge 2 + \frac{1}{2!} + \frac{1}{3!} + \cdots + \frac{1}{m!} = y_m.$$

Therefore, $y_m \le e$ for every m, and hence

$$f = \lim_{m \to \infty} y_m \le e. \tag{1.26}$$

We conclude from (1.25) and (1.26) that $f = e$, or, in other words,

$$e = \lim_{n \to \infty}\left(1 + \frac{1}{1!} + \frac{1}{2!} + \cdots + \frac{1}{n!}\right). \tag{1.27}$$

(In Chapter 2 we will write formulas like (1.27) in terms of "infinite series," in this case $e = \sum_0^\infty 1/n!$)

Formula (1.27) leads to an efficient scheme for numerical calculation of e to any desired degree of accuracy. Let us show first that for $n \geq 1$,

$$y_n < e \leq y_n + \frac{1}{n \cdot n!}. \tag{1.28}$$

If n is a given integer, let $m > n$. Then

$$y_m = y_n + \frac{1}{(n+1)!} + \cdots + \frac{1}{m!}$$

$$= y_n + \frac{1}{(n+1)!}\left[1 + \frac{1}{n+2} + \frac{1}{(n+2)(n+3)} + \cdots + \frac{1}{(n+2)\cdots m}\right]$$

$$< y_n + \frac{1}{(n+1)!}\left[1 + \frac{1}{n+1} + \frac{1}{(n+1)^2} + \cdots + \frac{1}{(n+1)^{m-n-1}}\right]$$

$$< y_n + \frac{1}{(n+1)!}\frac{1}{1 - \dfrac{1}{n+1}} = y_n + \frac{1}{n \cdot n!},$$

where we have again used Formula (1.19) for the sum of a geometric progression. Since $y_m < y_n + 1/(n \cdot n!)$ for every m, we see that

$$e = \lim_{m \to \infty} y_m \leq y_n + \frac{1}{n \cdot n!},$$

which proves (1.28).

On the basis of formula (1.28) we can devise an "automatic" scheme for computing e to any prescribed accuracy. Define two sequences $\{y_n\}$ and $\{\delta_n\}$ recursively as follows:

$$y_0 = 1, \qquad\qquad \delta_0 = 1$$

$$y_1 = y_0 + \delta_0, \qquad \delta_1 = \frac{\delta_0}{2}$$

$$y_2 = y_1 + \delta_1, \qquad \delta_2 = \frac{\delta_1}{3}$$

$$\cdot$$
$$\cdot$$
$$\cdot$$

$$y_n = y_{n-1} + \delta_{n-1}, \quad \delta_n = \frac{\delta_{n-1}}{n+1}.$$

Table I Numerical Calculation of e

n	y_n	δ_n	ε_n is less than:
0	1.000 000 0	1.000 000 0	1.0
1	2.000 000 0	0.500 000 0	0.3
2	2.500 000 0	0.166 666 7	0.06
3	2.666 666 7	0.041 666 7	0.02
4	2.708 333 4	0.008 333 3	0.003
5	2.716 666 7	0.001 388 8	0.0003
6	2.718 055 5	0.000 198 4	0.00003
7	2.718 253 9	0.000 024 8	0.000004
8	2.718 278 7		

It is easy to check that $\delta_n = 1/(n+1)!$, and therefore

$$y_n = 1 + 1 + \frac{1}{2!} + \cdots + \frac{1}{n!},$$

as before. We also define

$$\varepsilon_n = \frac{\delta_n}{n+1};$$

then by (1.28)

$$y_n < e < y_n + \varepsilon_{n-1}.$$

Table I shows a sample calculation of e, with an error $\leq 10^{-5}$, using the above formulas to calculate y_n, δ_n, and ε_n in order. The result at stage 8 is: $y_8 < e < y_8 + \varepsilon_7$; that is,

$$2.718\ 278\ 7 < e < 2.718\ 282\ 7.$$

Thus $e = 2.71828$ to within an error $< 10^{-5}$.† Table I was calculated by hand in less than 10 minutes. Similar methods are known for the computation of values of the functions e^x, $\sin x$, and $\cos x$. (Such methods can also be used in automatic digital computers, but they are *not* really used in practice, because much more efficient methods have been invented! By the way, a *computer program* consists essentially of just two items: (a) a recurrence scheme for calculating values of a sequence (or several sequences), and (b) instructions on where to stop.)

---

† This does not take into account the possibility of accumulated roundoff errors, which however cannot exceed 8×10^{-7}. Therefore, we are certain that the computed value of e is correct to within 10^{-5}.

Exercises

1. Calculate e to within 10^{-6}. (Use Table I. What about using Table I for an error $<10^{-7}$?)

2. (a) Show by induction that

$$\frac{2^n}{n!} < \frac{1}{2^{n-4}} \quad \text{if } n \geq 4.$$

   (b) Prove that $\lim\limits_{n \to \infty} \left(1 + \frac{2}{n}\right)^n$ exists.

3. Prove that

$$\lim_{n \to \infty} \left(1 + \frac{2}{n}\right)^n = \lim_{n \to \infty} \left(1 + 2 + \frac{2^2}{2!} + \cdots + \frac{2^n}{n!}\right).$$

4. Prove that $\lim_{n \to \infty} \left(1 + \frac{2}{n}\right)^n = e^2$. (*Hint:* note that $(1 + 2/n)^n = [(1 + 1/(n/2))^{n/2}]^2$.)

1.10   Growth Properties of Certain Sequences

The following notation is quite convenient.

Definition   *We say that a given sequence $\{x_n\}$ **diverges to** $+\infty$, (notation $\lim_{n \to \infty} x_n = +\infty$) if,*

for any given positive number M, there is an integer N (depending on M) such that $x_n > M$ for every $n \geq N$.

The sequences $\{n^2\}$ and $\{2^n\}$ are obvious examples of sequences diverging to $+\infty$.

In order to be consistent we will continue to say that the limit of $\{x_n\}$ *does not exist* if $\lim_{n \to \infty} x_n = +\infty$.

Theorem 1   *Let $\{x_n\}$ be a sequence of positive numbers. Then*

$$\lim_{n \to \infty} x_n = +\infty \quad \text{if and only if} \quad \lim_{n \to \infty} \frac{1}{x_n} = 0.$$

Proof Suppose $x_n \to +\infty$. Then for any $\varepsilon > 0$ we can determine an integer N such that $x_n > 1/\varepsilon$ when $n \geq N$. Therefore, $1/x_n < \varepsilon$ for $n \geq N$, which implies (since $1/x_n > 0$) that $1/x_n \to 0$. The converse is similar. ∎

Lemma   *Let $\{x_n\}$ be a sequence of nonzero numbers. If*

$$\lim_{n \to \infty} \left| \frac{x_{n+1}}{x_n} \right| < 1 \tag{1.29}$$

then $\lim_{n \to \infty} x_n = 0$.

Proof Let the limit in Equation (1.29) be denoted by B, and choose C with $B < C < 1$. There exists an integer N such that

$$|x_{n+1}| < C|x_n| \quad \text{for all } n \geq N.$$

By induction we then see that

$$|x_{N+k}| < C^k |x_N| \quad \text{for } k = 1, 2, 3, \ldots .$$

Now we know (Section 1.7, Theorem 3) that $C^k \to 0$ as $k \to \infty$. Hence given $\varepsilon > 0$ there exists an integer N_1 such that $C^k < \varepsilon/|x_N|$ for $k > N_1$. Thus for $n > N + N_1$ we have

$$|x_n| < \varepsilon. \quad ∎$$

It is a well-known fact (which has important applications in physics, chemistry, biology, economics, and other subjects) that any exponential function a^x (with $a > 1$) grows more rapidly as $x \to +\infty$ than any power of x. In terms of sequences this can be expressed as follows.

Theorem 2   *Let $a > 1$ and $p > 0$ be given. Then*

$$\lim_{n \to \infty} \frac{a^n}{n^p} = +\infty. \tag{1.30}$$

Proof Write $x_n = n^p/a^n$ and apply the above lemma:

$$\lim_{n \to \infty} \frac{x_{n+1}}{x_n} = \lim_{n \to \infty} \frac{1}{a} \cdot \left(\frac{n+1}{n} \right)^p = \frac{1}{a}.$$

Since $1/a < 1$ we conclude that $\lim_{n \to \infty} x_n = 0$, and Theorem 1 then implies that (1.30) holds. ∎

The following notation is often encountered.

Definition   *The expression*

$$x_n = o(y_n) \quad \text{as } n \to \infty$$

(read: "x_n is little-oh of y_n") means that

$$\lim_{n \to \infty} \frac{x_n}{y_n} = 0.$$

Similarly

$$x_n = O(y_n) \quad \text{as } n \to \infty$$

("x_n is big-oh of y_n") means that $\{x_n/y_n\}$ is a bounded sequence.

If $x_n = o(y_n)$ we also say that the sequence $\{x_n\}$ is *of smaller order than* $\{y_n\}$. Theorem 2 can therefore be interpreted as saying that $\{n^p\}$ is of smaller order than $\{a^n\}$ (given $a > 1, p > 0$). Other examples are given in the Exercises.

Exercises

1. The following sequences $\{x_n\}$ approach $+\infty$ as $n \to \infty$. Determine for each an integer N such that

    $$x_n > 1{,}000 \quad \text{for all } n \geq N.$$

    (a) $x_n = \dfrac{n^2 - 2}{2n + 3}$;   (b) $x_n = \sqrt[3]{n - 6}$;   (c) $x_n = \left(\dfrac{3}{2}\right)^n$.

2. Show that $\lim_{n \to \infty} \dfrac{a^n}{n!} = 0$. (What restrictions, if any, are required on a?)

3. Show that $\lim_{n \to \infty} \dfrac{n^n}{n!} = \infty$. Express the results of Exercises 2 and 3 in the "little-oh" notation.

4. Assume $a > 0$. Determine $\lim_{n \to \infty} \{(a+1)^n - a^n\}$; give reasons for your conclusion. (*Hint:* Consider the cases $a < 1$ and $a \geq 1$ separately.)

5. (a) Express the statement

    $$x_n = O(1) \quad \text{as } n \to \infty$$

    in terms of ε, N language. What does this say about the sequence $\{x_n\}$?

    (b) Same question for $x_n = o(1)$ as $n \to \infty$.

6. Show that if $x_n = o(1)$ as $n \to \infty$ and $y_n = O(1)$ as $n \to \infty$, then $x_n y_n = o(1)$ as $n \to \infty$. Express this in everyday language.

7. Suppose that $\lim_{n \to \infty} x_n = +\infty$ and $\lim_{n \to \infty} y_n = a > 0$. Prove that

    $$\lim_{n \to \infty} x_n y_n = +\infty.$$

8. Given that $\lim_{n \to \infty} x_n = +\infty$, prove that

    $$\lim_{n \to \infty} \frac{x_n}{1 + x_n} = 1.$$

    Also prove the converse under the additional assumption that $x_n > 0$ for all n.

9. Let k be a fixed integer and let $a_1, a_2, \ldots, a_k$ be constants. Prove that

    $$\lim_{n \to \infty} (n^k + a_1 n^{k-1} + a_2 n^{k-2} + \cdots + a_k) = +\infty.$$

    Do not assume $a_i \geq 0$. (*Hint:* Use Exercise 7.)

10. If $b > 1$, show that

    $$\lim_{n \to \infty} \frac{b^n}{n^k + a_1 n^{k-1} + a_2 n^{k-2} + \cdots + a_k} = +\infty.$$

1.11   Some Further Properties of the Real Number System

In this chapter we have stressed a fundamental distinction between the system $\mathbb{Q}$ of all rational numbers, and the system $\mathbb{R}$ of all real numbers, namely the property of completeness. We have indicated the importance of this property in the study of limits. In this section we wish to discuss briefly some questions related to the fact that the number systems $\mathbb{Q}$ and $\mathbb{R}$ (as well as the system $\mathbb{N}$ of all positive integers) contain infinitely many elements. Questions about the infinite have plagued mathematicians, philosophers, and students since classical times. Since the pioneering work of Georg Cantor, however, mathematicians have been able to develop virtually complete mastery over the concept of the infinite.

Cantor's brilliant contribution to mathematics begins with the following definition, which tells us under what conditions two sets (finite or infinite) have the "same number" of elements. Since "number" already had a fairly definite meaning in mathematics, Cantor introduced the term "equivalent" sets.

Definition 1   *Two sets A and B are said to be **equivalent** if there exists a one-to-one correspondence between the elements of A and the elements of B.*

Obviously two finite sets A and B are equivalent if and only if they have the same number of elements, in the usual sense. Cantor's definition is most interesting, however, when applied to infinite sets.

For example, if $\mathbb{N} = \{1, 2, 3, \ldots\}$ is the set of positive integers, let $\mathbb{N}_2 = \{2, 4, 6, \ldots\}$ be the set of all even positive integers. These sets are equivalent, since the correspondence $n \leftrightarrow 2n$ is a one-to-one correspondence between the elements of $\mathbb{N}$ and $\mathbb{N}_2$.

Note that $\mathbb{N}_2$ is a proper subset of $\mathbb{N}$. Thus an *infinite* set may be equivalent to a proper subset of itself; in fact, this property is characteristic of infinite sets (see Exercise 4).

Any set which is equivalent to the set $\mathbb{N}$ of positive integers is said to be *countable*, or *countably infinite* (or, sometimes, *enumerable*). *Thus the set $\mathbb{N}_2$ of* even integers is countable; it is easy to show (Exercise 1) that the set $\mathbb{Z}$ of all integers, positive and negative, is also countable. The question immediately arises whether all infinite sets are countable, and in particular, what about the sets $\mathbb{Q}$ and $\mathbb{R}$?

Theorem 1 (Cantor)   *The set $\mathbb{Q}$ of all rational numbers is countable, but the set $\mathbb{R}$ of all real numbers is uncountable.*

Proof A given set S is countable if and only if there is a one-to-one function from $\mathbb{N}$ to S, such that each member of S corresponds to a unique positive integer n. But a function from $\mathbb{N}$ to S is nothing but a sequence. Hence we can say that S is countable if and only if there exists a sequence $\{x_n\}$ such that every member of S occurs exactly once as a term of the sequence.

(a) $\mathbb{Q}$ is countable: Consider the following sequence $\{x_n\}$,

$$\underbrace{\tfrac{0}{1},}_{(1)} \; \underbrace{\tfrac{1}{1},}_{(2)} \; \underbrace{-\tfrac{1}{1}, \tfrac{1}{2},}_{(3)} \; \underbrace{-\tfrac{1}{2}, \tfrac{2}{1}, -\tfrac{2}{1}, \tfrac{1}{3},}_{(4)} \; -\tfrac{1}{3}, \tfrac{3}{1}, -\tfrac{3}{1}, \dots,$$

where the kth group contains all positive and negative rational numbers in lowest terms, whose numerator and denominator add up to k. Clearly every rational number p/q occurs exactly once in this sequence.

(b) $\mathbb{R}$ is uncountable: Suppose to the contrary that $\mathbb{R}$ is countable, and let $\{x_n\}$ be a sequence containing all real numbers. Let us write out the decimal expansions of each x_n:

$$\begin{aligned} x_1 &= n_1 \,.\, a_{11}a_{12}a_{13} \cdots \\ x_2 &= n_2 \,.\, a_{21}a_{22}a_{23} \cdots \\ x_3 &= n_3 \,.\, a_{31}a_{32}a_{33} \cdots \end{aligned} \tag{1.31}$$

and so forth. Here the numbers a_{ij} are digits (0 to 9). According to our assumption, *every* real number appears on this list somewhere.

Now for each $i = 1, 2, 3, \dots$, let b_i be any digit except a_{ii}, 0, or 9. Consider the real number

$$x = 0 \,.\, b_1 b_2 b_3 \dots.$$

Since the decimal expansion of x is different from every decimal in the sequence (1.31) (and since x did not arise through the ambiguity of repeating 9's), we reach the contradiction that x is *not* on the list (1.31). This completes the proof. This method of proving (b) is sometimes called "Cantor's diagonalization process." ∎

The proof of the following theorem, which is a generalization of Theorem 1, is left for the exercises.

Theorem 2   Let $a < b$ *be two arbitrary real numbers. Then the set of all rational numbers lying between a and b is (infinite and) countable, whereas the set of all irrational numbers lying between a and b is (infinite and) uncountable.*

Definition 2   *Any set S with the property that every open interval (a, b), $a < b$, contains at least one point of S, is said to be a **dense subset** of $\mathbb{R}$ (see Figure 1.4).*

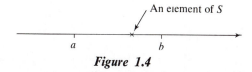

Figure 1.4

We see, for example, by Theorem 2, that $\mathbb{Q}$ is a countable, dense subset of $\mathbb{R}$.

Theorem 3   *If S is a dense subset of* $\mathbb{R}$*, then every real number r is the limit of some sequence* $\{x_n\}$ *of members of S.*

Proof   For each positive integer n, consider the open interval $(r - 1/n, r + 1/n)$. By hypothesis, there is a member x_n of S in this interval. Thus $|x_n - r| < 1/n$, and therefore, $\lim_{n \to \infty} x_n = r$.   ∎

Exercises

1. Show that the set $\mathbb{Z} = \{0, \pm 1, \pm 2, \ldots\}$ of all integers is countable.

2. Let W denote the set of all ordered pairs (n, m) of positive integers. Show that W is countable.

3. Let S and T be countable sets. Show that the *union* $S \cup T$ (consisting of all members of S together with all members of T) is also countable.

*4. Let A be an infinite set, and let B be obtained by deleting one element of A. Show that A and B are equivalent.

5. Show that any interval (a, b), $a < b$, is equivalent to the interval $(-1, +1)$. Hence, all nonempty, open intervals are equivalent.

6. By means of the function $f(x) = x/\sqrt{x^2 + 1}$, show that the interval $(-1, +1)$ is equivalent to $\mathbb{R}$ (see Figure 1.5).

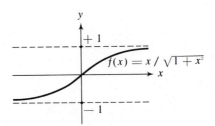

Figure 1.5

7. Prove Theorem 2.

8. Let S be a set derived from $\mathbb{Q}$ by removing finitely many elements of $\mathbb{Q}$. Show that S is a dense subset of $\mathbb{R}$. Could infinitely many elements be removed in some way, still leaving a dense subset of $\mathbb{R}$?

9. Which of the following subsets are dense in $\mathbb{R}$?
   (a) a finite set $\mathbb{F}$;
   (b) the set $\mathbb{Z}$ of all integers;
   (c) the set of all rational numbers with even denominators (in lowest terms);
   (d) the set of "dyadic" rationals $\dfrac{n}{2^m}$, $n \in \mathbb{Z}$, $m \in \mathbb{N}$.

2 *Infinite Series*

2.1 Elementary Properties

In this chapter we develop the elementary theory of infinite series of real numbers, using some of the results obtained in Chapter 1.

Let $\{a_n\}$ be a given sequence of real numbers. We can define a new sequence $\{s_n\}$ by forming the sums

$$s_n = \sum_{k=1}^{n} a_k = a_1 + a_2 + \cdots + a_n. \tag{2.1}$$

If it happens that the new sequence $\{s_n\}$ converges to the limit S ($\lim_{n\to\infty} s_n = S$), we write

$$\sum_{n=1}^{\infty} a_n = S, \tag{2.2}$$

and we call S the *sum of the infinite series* $\sum_{n=1}^{\infty} a_n$. The sum s_n of (2.1) is called the *n*th *partial sum* of the series $\sum_{n=1}^{\infty} a_n$.

In other words, by definition

$$\sum_{n=1}^{\infty} a_n = \lim_{n\to\infty} \sum_{k=1}^{n} a_k, \tag{2.3}$$

provided this limit exists. If the limit does exist, we say that the infinite series $\sum_{n=1}^{\infty} a_n$ *converges* (to S). If $\lim_{n\to\infty} s_n$ does not exist, we say that $\sum_{n=1}^{\infty} a_n$ *diverges*. In the special case that $\lim_{n\to\infty} s_n = +\infty$ we say that $\sum_{n=1}^{\infty} a_n$ *diverges to* $+\infty$.

Example 1

$\sum_{n=1}^{\infty} \dfrac{1}{2^n} = 1$.   To verify this we calculate the partial sums:

$$s_n = \sum_{k=1}^{n} \frac{1}{2^k} = \frac{1}{2} + \frac{1}{2^2} + \cdots + \frac{1}{2^n}$$

$$= 1 - \frac{1}{2^n},$$

by the formula for summing a geometric progression. It follows that $\lim_{n \to \infty} s_n = 1$, which means the same as $\sum_{n=1}^{\infty} 1/2^n = 1$.

Example 1 will be generalized in Theorem 2 below.

Example 2

$\sum_{n=1}^{\infty} (-1)^n$ diverges, because

$$s_1 = -1,$$
$$s_2 = -1 + 1 = 0,$$
$$s_3 = -1 + 1 - 1 = -1,$$

and so forth. Thus the sequence $\{s_n\}$ is

$$-1, 0, -1, 0, -1, 0, \ldots,$$

which surely diverges.

Example 3

$\sum_{n=1}^{\infty} \dfrac{n}{n+1}$ diverges to $+\infty$.

To check this we notice first that $\dfrac{n}{n+1} \geq \frac{1}{2}$ (by algebra) for all n; there-fore,

$$s_n = \frac{1}{2} + \frac{2}{3} + \cdots + \frac{n}{n+1} > n \cdot \frac{1}{2};$$

so $s_n \to +\infty$ as $n \to \infty$.

It seems evident that a series $\sum_1^\infty a_n$ can't converge unless $a_n \to 0$ as $n \to \infty$. We can easily prove that this is the case.

Theorem 1   *If $\sum_1^\infty a_n$ converges, then $\lim_{n\to\infty} a_n = 0$. The converse is false.*

Proof Suppose first that $\sum_1^\infty a_n$ converges. Then $\lim_{n\to\infty} s_n = s$ exists. Note that

$$a_{n+1} = s_{n+1} - s_n.$$

It follows that

$$\lim_{n\to\infty} a_{n+1} = \lim_{n\to\infty} s_{n+1} - \lim_{n\to\infty} s_n = s - s = 0.$$

Therefore also $\lim_{n\to\infty} a_n = 0$.

To prove that the converse is false, we consider the following case, known as the harmonic series, in which $a_n \to 0$ but $\sum_1^\infty a_n$ diverges to $+\infty$. ▮

Example 4

The harmonic series $\displaystyle\sum_{n=1}^\infty \frac{1}{n}$ diverges to $+\infty$.   (2.4)

We can establish this by using simple calculus. (A proof not using calculus is described in Exercise 7.) Consider the curve

$$y = f(x) = \frac{1}{x} \qquad (x > 0).$$

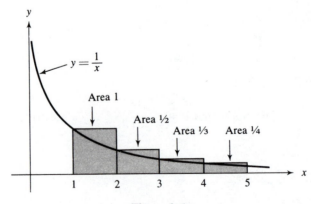

Figure 2.1

Since $f(x) = 1/x$ is decreasing, it is fairly obvious from Figure 2.1 that we have

$$s_n = 1 + \frac{1}{2} + \frac{1}{3} + \cdots + \frac{1}{n} > \int_1^{n+1} \frac{1}{x}\, dx = \log(n+1).$$

Hence, $s_n \to +\infty$ as $n \to \infty$, which means $\sum_1^\infty \frac{1}{n}$ diverges to $+\infty$. ∎

The simple method above is a special case of the "integral test," discussed in Section 2. The same method can also be used to prove that certain series converge.

Example 5

$\sum_{n=1}^\infty \frac{1}{n^2}$ converges.

To verify this, consider the graph of $y = f(x) = 1/x^2$ in Figure 2.2.

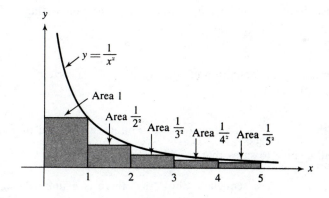

Figure 2.2

We see that

$$s_n = \sum_{k=1}^n \frac{1}{k^2} < 1 + \int_1^n \frac{dx}{x^2} = 2 - \frac{1}{n} < 2.\dagger \qquad (2.5)$$

---

† We recall that the inequality (2.5) was to be proved by induction in Exercise 4 of Section 1.8.

Hence, $\{s_n\}$ is a bounded, increasing sequence, which must therefore converge, by the completeness property.†

The method used in Examples 4 and 5 can also be used to derive the following general result, as the reader can readily verify.

Theorem 2 (*The hyperharmonic series*)

$$\sum_{n=1}^{\infty} \frac{1}{n^p} \quad converges \quad if \, and \, only \, if \quad p > 1. \qquad (2.6)$$

Exercises

1. Find expressions for the partial sums $s_1, s_2, \ldots, s_5$ for the series:

   (a) $\displaystyle\sum_{n=1}^{\infty} \frac{1}{2n+1}$, (b) $\displaystyle\sum_{n=1}^{\infty} \frac{1}{(n^2)!}$.

   If you have a calculator, evaluate these partial sums to, say, 5 decimals. Can you give an argument for the convergence or divergence of these series? (Routine tests will be given in the next section.)

2. Suppose that $\sum_{1}^{\infty} a_n$ and $\sum_{1}^{\infty} b_n$ both converge, and prove that

   $$\sum_{1}^{\infty} (\alpha a_n + \beta b_n) = \alpha \sum_{1}^{\infty} a_n + \beta \sum_{1}^{\infty} b_n.$$

   (You must use the definition of convergence, plus a certain theorem about sequences.)

3. Show that $\displaystyle\sum_{n=2}^{\infty} \frac{1}{n \log n}$ diverges to $+\infty$. Follow the method of Example 4; you will be able to show that $s_n > \log [\log (n+1)]$.

4. Let $\{s_n\}$ be a *given* sequence. Show that there is a uniquely determined sequence $\{a_n\}$ such that $\{s_n\}$ is the sequence of partial sums of the series $\sum_{1}^{\infty} a_n$. This shows (surprisingly) that the theories of sequences and of series are logically equivalent! (*Hint:* $s_1 = a_1$; $s_2 = a_1 + a_2$; and so forth. Solve for $a_1, a_2, \ldots$.)

5. Evaluate $\sum_{0}^{\infty} 1/n!$.

---

† It can be shown by methods of advanced calculus that $\displaystyle\sum_{1}^{\infty} \frac{1}{n^2} = \frac{\pi^2}{6} \cong 1.64$.

6.   Prove that $\displaystyle\sum_{n=1}^{\infty} \frac{1}{n(n+1)} = 1$. (Try to guess a simple formula for s_n, and then prove it.)

7.   Prove by induction that

$$\sum_{k=1}^{2^n} \frac{1}{k} = 1 + \frac{1}{2} + \frac{1}{3} + \cdots + \frac{1}{2^n} > \frac{n}{2}.$$

Hence, the series $\displaystyle\sum_{1}^{\infty} \frac{1}{n}$ can be shown to diverge without using integrals or logarithms.

2.2   Convergence Tests

In this section we establish a variety of tests for the convergence of series, and show how they are used. The student should try to develop the ability to *guess* whether a given series converges; the appropriate test will often become clear during the process of formulating a guess.

Lemma 1   *Let $\sum_{1}^{\infty} a_n$ be a series of nonnegative terms (that is, $a_n \geq 0$ for all n). Then the series converges if and only if the sequence $\{s_n\}$ of partial sums is bounded.*

Proof   By definition $\sum_{1}^{\infty} a_n$ converges if and only if the sequence $\{s_n\}$ converges. But $a_n \geq 0$ implies $s_n \geq s_{n-1}$, or, in other words, $\{s_n\}$ is *nondecreasing*. By the completeness property, such a sequence converges if and only if it is bounded.   ∎

Of course this lemma does not hold if the terms a_n are of arbitrary sign (why not?).

The following theorem is simple but important.

Theorem 1 (A comparison test for convergence)   *Let $\sum_{1}^{\infty} b_n$ be a convergent series of positive terms. Let $\sum_{1}^{\infty} a_n$ be another series, such that*

$$0 \leq a_n \leq Mb_n \quad \text{for all } n \ (M = \text{constant}). \tag{2.7}$$

Then $\sum_{1}^{\infty} a_n$ also converges.

Proof   The hypotheses clearly imply that the partial sums $s_n = \sum_{k=1}^{n} a_k$ are bounded and nondecreasing. Hence $\sum_{1}^{\infty} a_n$ converges.   ∎

In using this comparison test, it is often useful to realize that if (2.7) is known to hold for all sufficiently large n, then it necessarily holds for all n (with perhaps a larger M).

Corollary (A comparison test for divergence)   *Let $\sum_{1}^{\infty} c_n$ be a divergent series of positive terms. Let $\sum_{1}^{\infty} a_n$ be another series such that*

$$a_n \geq mc_n \quad \text{for all } n \quad (m = \text{constant} > 0).$$

Then $\sum_{1}^{\infty} a_n$ diverges.

The proof is trivial.

Example 1

$$\sum_{n=1}^{\infty} \frac{6n}{n^4 + 3} \text{ converges.}$$

To see this, note that $6n/(n^4 + 3) < 6/n^3$. Since the hyperharmonic series $\sum_{1}^{\infty} 1/n^3$ converges, the given series must also converge.

Another simple series which is often used for comparison is the following:

Theorem 2 (The geometric series)

$$\sum_{n=0}^{\infty} a^n = \frac{1}{1 - a} \quad \text{if } |a| < 1. \tag{2.8}$$

The series diverges if $|a| \geq 1$.

Proof   We know from Section 1.7, Formula (1.20), that

$$s_n = 1 + a + a^2 + \cdots + a^n = \frac{1 - a^{n+1}}{1 - a} \rightarrow \frac{1}{1 - a} \quad \text{if } |a| < 1,$$

and this is the same as (2.8) above. Suppose on the other hand that $|a| \geq 1$. Then $\lim_{n \to \infty} a^n \neq 0$, so that the series $\sum_0^\infty a^n$ must diverge. ∎

In using comparison tests, the following simple observation is useful.

Lemma 2   Let $\{a_n\}$ and $\{b_n\}$ be two sequences of positive numbers. If $\lim_{n \to \infty} a_n/b_n$ exists, then there exists a constant M such that

$$a_n \leq M b_n \quad \text{for all } n.$$

Proof   Any convergent sequence is automatically bounded, so we know that

$$\frac{a_n}{b_n} \leq M = \text{constant} \quad \text{for all } n. \quad ∎$$

Theorem 3   Let $\sum a_n$ and $\sum b_n$ be series of positive terms. Assume that $\lim_{n \to \infty} a_n/b_n = L$ exists. Then

(a) if $\sum b_n$ converges, so does $\sum a_n$;
(b) if $\sum b_n$ diverges and $L > 0$, then $\sum a_n$ diverges.

Proof   Part (a) is obvious from Lemma 2. For (b), note that if $L > 0$, then

$$\frac{a_n}{b_n} > \tfrac{1}{2}L > 0$$

for large n. Hence, $\sum a_n$ must diverge if $\sum b_n$ does. ∎

If a given series $\sum a_n$ having all *positive* terms is encountered, it is frequently (but not always!) possible to determine whether the series converges or diverges by "common sense" reasoning, based on the above simple theorems. First, if a_n does not approach zero as $n \to \infty$, the series diverges. But more precisely, even if $a_n \to 0$ as $n \to \infty$, the series will diverge unless $a_n \to 0$ "faster than" $1/n$. Conversely, if $a_n \to 0$ "faster than" $1/n^p$ for some $p > 1$, then the series must converge. Difficult cases arise when a_n approaches zero at approximately the same rate as $1/n$, or when a_n behaves irregularly.

Example 2

The series

$$\sum_1^\infty \frac{n^3 - 6}{3(n^2 + 2n - 1)(n^2 + 5)}$$

diverges, because the nth term behaves approximately like $1/3n$ (why?), and the series $\sum 1/3n$ is known to diverge. This can be proved rigorously by using Theorem 3.

For certain series the following convergence test is easily applied.

Theorem 4 (The ratio test)   *Let $\sum_1^\infty a_n$ be a series of positive terms, and assume that the ratios of successive terms converge to a limit L, i.e.*

$$\lim_{n \to \infty} \frac{a_{n+1}}{a_n} = L. \tag{2.9}$$

Then the series $\sum_1^\infty a_n$

(a) *converges   if $L < 1$,*

(b) *diverges   if $L > 1$.*

(Note that no conclusion can be drawn in the event that $L = 1$; alternative methods must be used to check for convergence in such cases.)

The proof of Theorem 4 will be covered in the Exercises.

Example 3

The series $\sum_1^\infty na^n$ converges if $0 < a < 1$ and diverges if $a \geq 1$.

To see this, consider the ratio

$$\frac{a_{n+1}}{a_n} = a \cdot \frac{n+1}{n} \to a \quad \text{as} \quad n \to \infty.$$

Hence the result follows immediately from the ratio test, except when $a = 1$, for which case the series obviously diverges.

Theorem 5 (The integral test)   *Let $f(x)$ be a continuous positive decreasing function defined for all $x \geq 1$. Let*

$$a_n = f(n).$$

Then the series $\sum_1^\infty a_n$ converges if and only if the improper integral $\int_1^\infty f(x)\,dx$ converges.

(This improper integral converges, by definition, if $\lim_{b \to \infty} \int_1^b f(x)\,dx$ exists. See Section 5.4 for the theory of improper integrals.)

Proof The proof follows the same lines as Examples 4 and 5 of the previous section. For example, assume that $\int_1^\infty f(x)\,dx$ converges. Then, as in Figure 2.2 we have

$$s_n = \sum_{k=1}^n a_k \le a_1 + \int_1^n f(x)\,dx \le a_1 + \int_1^\infty f(x)\,dx,$$

the last expression being a (finite) real number, M. Thus the partial sums s_n are bounded, so that the series $\sum_1^\infty a_n$ converges, by Lemma 1.

The converse is proved similarly. ∎

Theorem 6 (The root test) Let $\sum_1^\infty a_n$ *be a series of positive terms, and assume that the nth root of a_n approaches a limit α, i.e.*

$$\alpha = \lim_{n \to \infty} \sqrt[n]{a_n}. \tag{2.10}$$

Then the series $\sum_1^\infty a_n$

(a) *converges* *if $\alpha < 1$,*

(b) *diverges*   *if $\alpha > 1$.*

(No conclusion can be drawn if $\alpha = 1$.)

Proof Suppose first that $\alpha < 1$, and choose β such that $\alpha < \beta < 1$. Then for large n we have

$$\sqrt[n]{a_n} < \beta, \qquad \text{or} \qquad a_n < \beta^n.$$

Since the geometric series $\sum_1^\infty \beta^n$ converges, so must the given series $\sum_1^\infty a_n$, by comparison.

The converse is left to the reader. ∎

Now, how to use these tests? As noted at the end of Section 1, convergence of a positive series $\sum_1^\infty a_n$ depends on the *speed* with which $a_n \to 0$ (roughly speaking). For some examples, it is easily seen that $a_n = O(1/n^p)$ for some $p > 0$; if $p > 1$ the series converges by comparison with $\sum_1^\infty 1/n^p$. Other examples are seen to diverge by similar logic. On the other hand, many series are "geometric" in nature, e.g. $a_n = O(\beta^n)$ for some $\beta > 0$. If $\beta < 1$ the series converges (why?). In the latter case, the ratio (or root) test often provides the fastest test for convergence. The following exercises should make this clear. (Nonobvious examples do sometimes arise!)

Exercises

1. Prove the convergence or divergence of each of the following series. Unless otherwise indicated, all sums run from $n = 1$ to $n = \infty$.

    (a) $\sum \dfrac{n-1}{2n-1}$

    (b) $\sum \dfrac{n}{n^4+1}$

    (c) $\sum \dfrac{1}{\sqrt{n(n+1)(n+2)}}$

    (d) $\displaystyle\sum_{n=2}^{\infty} \dfrac{1}{n(\log n)^2}$

    (e) $\sum \dfrac{\log n}{n^2}$

    (f) $\sum \dfrac{n^3}{2^n}$

    (g) $\sum \dfrac{3^n}{n\cdot 4^n}$

    (h) $\sum \dfrac{2^n}{(n!)^2}$

    (i) $\displaystyle\sum_{n=2}^{\infty} \dfrac{1}{(\log n)^n}$

    (j) $\sum n^{-\sqrt{n}}$

    (k) $\sum \sin \dfrac{\pi}{n}$

    (l) $\sum \dfrac{(2n)!}{(n!)^2}$

    (m) $\sum \dfrac{\sqrt{n+1}-\sqrt{n}}{n}$

    (n) $\sum \dfrac{1}{2+\sin n}$

2. Prove that the hyperharmonic series $\sum 1/n^p$ converges if and only if $p > 1$.

3. Let $\{a_n\}$ be a sequence of positive numbers. Show that if $\sum a_n$ converges, then $\sum a_n^2$ also converges, but not conversely.

*4. If $\sum a_n$ diverges, and $a_n > 0$ for all n, show that $\sum \dfrac{a_n}{1+a_n}$ diverges. (*Hint:* Argue by contradiction!)

5. Prove part (a) of Theorem 4. (*Hint:* Show from Equation (2.9) that there is an integer N such that

    $$a_{N+k} \le \beta^k a_N \quad \text{for} \quad k = 1, 2, 3, \dots$$

    where β is some appropriate number, $0 < \beta < 1$; cf. the lemma of Section 1.10.)

6. Prove part (b) of Theorem 4.

7. Find examples of both a convergent series and a divergent series for which the limit L in Equation (2.9) equals 1.

8. If $\sum a_n$ and $\sum b_n$ are convergent series of positive terms, does it follow that $\sum \sqrt{a_n^2 + b_n^2}$ converges? Conversely?

2.3   Series with Positive and Negative Terms

Most of the discussion so far has dealt with series $\sum a_n$ in which $a_n > 0$ for all n. We now drop the latter hypothesis.

Definition   *The series $\sum_{n=1}^{\infty} a_n$ **converges absolutely** if the series $\sum_{n=1}^{\infty} |a_n|$ is convergent. Any series that converges but does not converge absolutely is said to* **converge conditionally.**

Example 1

The series $\sum_{n=1}^{\infty} \dfrac{(-1)^n}{n}$ converges conditionally: we know that this series does not converge absolutely (why?); the fact that the series does converge will follow from Theorem 3 below.

Every absolutely convergent series also converges in the usual sense of the term, as we now show.†

Let $\sum a_n$ be absolutely convergent. Let s_n^+ and s_n^- denote respectively the sums of the positive and negative terms among $a_1, a_2, \ldots, a_n$. We then have

$$s_n^+ + s_n^- = s_n = \sum_{k=1}^{n} a_k \tag{2.11}$$

and

$$s_n^+ - s_n^- = S_n = \sum_{k=1}^{n} |a_k| \tag{2.12}$$

where s_n and S_n denote the partial sums indicated.

The sequence $\{s_n^+\}$ is nondecreasing (why?) and bounded:

$$s_n^+ \leq S_n \leq S = \sum_{1}^{\infty} |a_n|.$$

Similarly, the sequence $\{s_n^-\}$ is nonincreasing and bounded. By the completeness property the limits

$$s^+ = \lim_{n \to \infty} s_n^+ \qquad \text{and} \qquad s^- = \lim_{n \to \infty} s_n^- \tag{2.13}$$

---

† An alternative proof of this fact will be given later (see Exercise 4, Section 4.4.)

both exist. Consequently from Equation (2.11)

$$\lim_{n \to \infty} s_n = s^+ + s^-, \tag{2.14}$$

i.e. the original series $\sum_1^\infty a_n$ converges to the sum $s = s^+ + s^-$. This proves our result:

Theorem 1   If $\sum_1^\infty a_n$ converges absolutely, then it also converges in the usual sense.

An absolutely convergent series results from "sprinkling plus and minus signs" on the terms of a convergent series of positive terms

$$\left(\text{example: } 1 - \frac{1}{2^2} + \frac{1}{3^2} + \frac{1}{4^2} - \frac{1}{5^2} - \frac{1}{6^2} - \cdots\right).$$

Whenever this is done, the positive and negative "pieces" must themselves form convergent series—as the above proof tells us. Conversely, if the positive and negative pieces converge, the given series converges absolutely:

$$\lim_{n \to \infty} S_n = s^+ - s^-. \tag{2.15}$$

This implies that:

Theorem 2   If $\sum_1^\infty a_n$ converges conditionally then both the positive and negative "pieces" diverge, or in other words

$$\lim_{n \to \infty} s_n^+ = + \infty \quad \text{and} \quad \lim_{n \to \infty} s_n^- = - \infty. \tag{2.16}$$

The example given above, $\sum (-1)^n/n$, illustrates this result. However, we have not yet proved that this series actually converges! We now do so.

Definition   An infinite series $\sum a_n$ in which the terms alternate in sign is called an **alternating series**.

Theorem 3   Let $\sum_1^\infty a_n$ be an alternating series satisfying the conditions

$$0 < |a_{n+1}| < |a_n| \quad \text{for all } n \tag{2.17}$$

and

$$a_n \to 0 \quad as \; n \to \infty. \tag{2.18}$$

Then $\sum_1^\infty a_n$ converges.

Proof Suppose, to be explicit, that $a_1 > 0$. The facts that the signs of the a_i alternate, and that the a_i decrease in absolute value, means that the partial sums, $a_1, a_1 + a_2, a_1 + a_2 + a_3$, etc., "oscillate" as indicated in Figure 2.3. That is, we have

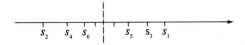

Figure 2.3

$$s_2 < s_4 < s_6 < \cdots < s_5 < s_3 < s_1. \tag{2.19}$$

For example, since $a_3 + a_4 = |a_3| - |a_4| > 0$ we obtain

$$s_4 = s_2 + a_3 + a_4 > s_2;$$

the other inequalities in (2.19) follow in the same way.

Next we employ the completeness property to conclude that the sequence $s_2, s_4, s_6, \ldots$ converges to a limit s, and also that $s_1, s_3, s_5, \ldots$ converges to a limit s'. Also, since

$$s_{2n+1} - s_{2n} = a_{2n+1} \to 0 \quad as \; n \to \infty$$

by (2.18), we see that $s = s'$. From this we finally conclude that $\lim_{n\to\infty} s_n = s$. (As an exercise in ε—N technique, prove this last statement in detail!) ∎

The simplest example of an alternating series that is conditionally convergent is the series $\sum (-1)^n/n$. Other examples appear in the exercises.

By a *rearrangement* of a given series $\sum a_n$ we mean another series, containing exactly the same terms as the given series, but in a different order. For example, the harmonic series $\sum 1/n$ could be rearranged to obtain

$$1 + \tfrac{1}{3} + \tfrac{1}{2} + \tfrac{1}{5} + \tfrac{1}{7} + \tfrac{1}{4} + \cdots.$$

It can be shown that

(a) if the series $\sum a_n$ converges absolutely to S, then every rearrangement also converges absolutely to S;

(b) if the series $\sum a_n$ converges conditionally to S, and if S' is any real number, then the series can be rearranged to converge to S'. It can even be rearranged to diverge.

Let us prove (b). Let $a_1^+, a_2^+, a_3^+, \ldots$ denote the positive terms, and $a_1^-, a_2^-, a_3^-, \ldots$ negative terms of the given conditionally convergent series. Since $\sum a_n$ converges we know that a_n^+ and a_n^- both $\to 0$ as $n \to \infty$. Also, from Theorem 2, both series $\sum a_n^+$ and $\sum(-a_n^-)$ diverge to $+\infty$.

The desired sum for the rearrangement is S'. Pick n_1 as the first integer such that

$$a_1^+ + a_2^+ + \cdots + a_{n_1}^+ > S'$$

Now pick n_2 as the first integer such that

$$(a_1^+ + a_2^+ + \cdots + a_{n_1}^+) + (a_1^- + a_2^- + \cdots + a_{n_2}^-) < S'.$$

Continue in this way, going alternately above and below S' using more and more positive and negative terms. This process can always be continued (why?), and the resulting sequence of a_i is a rearrangement of the original, which converges to S' because the terms a_n^+ and a_n^- go to zero. This prove the first assertion in (b). The second is left as an exercise.

Exercises

1. Test the following series for absolute or conditional convergence.

    (a) $\sum(-1)^n \log n$ (b) $\sum \dfrac{(-1)^n}{\sqrt{n}}$

    (c) $\sum \dfrac{(-1)^n}{n^2}$ (d) $\sum \dfrac{(-1)^{[n/2]}}{n}$

    (Note: $[x]$ denotes the greatest integer $\leq x$.)

    (e) $\sum(-n)^{-n}$ (f) $\sum \dfrac{\sin nx}{n^2}$

    (g) $\sum n(-\tfrac{1}{2})^n$ (h) $\sum \dfrac{(-1)^n + \tfrac{1}{2}}{n}$

2. Does the series

    $$1 - \frac{1}{2^2} + \frac{1}{3} - \frac{1}{4^2} + \frac{1}{5} - \cdots$$

    converge?

3. Does the series

$$1 + \tfrac{1}{2} - \tfrac{1}{3} + \tfrac{1}{4} + \tfrac{1}{5} - \tfrac{1}{6} + \cdots$$

(two plus signs followed by one minus) converge?

4. For what values of $a \in \mathbb{R}$ is the series $\sum a^n/n$ convergent? Absolutely convergent?

5. Try to think up a fairly simple looking series whose convergence will stump your professor.

6. The alternating harmonic series $\sum(-1)^n/n$ is to be rearranged to converge to zero. Write down the first 10 terms of the rearrangement described in the text (use a calculator).

7. Show that a conditionally convergent series can be rearranged to diverge to $+\infty$.

8. Prove that the sum of two absolutely convergent series converges absolutely. What about conditional convergence?

9. Can a conditionally convergent series be rearranged so as to converge absolutely? Why?

*10. Prove that the improper integral

$$\int_0^\infty \frac{\sin x}{x}\,dx$$

converges, i.e. that

$$\lim_{b \to \infty} \int_0^b \frac{\sin x}{x}\,dx$$

exists. (*Hint:* Draw a graph!)

2.4 *Power Series*

Definition   *A series of the form*

$$\sum_{n=0}^{\infty} a_n x^n \qquad\qquad (2.20)$$

*is called a **power series in** x; here the **coefficients** a_n are assumed given, and x denotes a real variable.*†

The theory of power series and their convergence will be discussed in detail in Chapter 5. Here we will prove only the following.

Theorem   *Assume that*

$$\alpha = \lim_{n \to \infty} \left| \frac{a_{n+1}}{a_n} \right| \tag{2.21}$$

exists, or that the limit is $+\infty$. Let $R = 1/\alpha$ if $\alpha \neq 0$; let $R = +\infty$ if $\alpha = 0$ and $R = 0$ if $\alpha = +\infty$.‡ Then the power series (2.20)

(a) *converges absolutely for all x with $|x| < R$,*

(b) *diverges for $|x| > R$.*

The number R is called the *radius of convergence* of the given power series.

Proof   Write $b_n = |a_n x^n|$; then

$$\frac{b_{n+1}}{b_n} = \left| \frac{a_{n+1}}{a_n} \right| |x|,$$

so that $\lim_{n \to \infty} |b_{n+1}/b_n| = \alpha|x|$. By Theorem 4 of Section 2, the series $\sum |a_n x^n| = \sum b_n$ converges if $\alpha|x| < 1$, i.e. if $|x| < 1/\alpha = R$, and this proves (a). Similarly, $\sum b_n$ diverges if $|x| > R$, proving (b). (The infinite cases are easily argued separately.)   ∎

Note that the above theorem does not specify the convergence of a power series at the end points $x = \pm R$ of the interval of convergence. In fact, anything may happen at these end points: the series may diverge, or converge conditionally or absolutely. Examples are given in the exercises.

---

† The value of (2.20) is initially undefined at $x = 0$, but we define its value at 0 to be simply a_0.
‡ The abuse of language here should not lead to confusion!

Example 1

Find the radius of convergence of the power series, $\sum_0^\infty x^n/n^2$.
    Here we have $a_n = 1/n^2$ so that

$$\alpha = \lim_{n \to \infty} \left| \frac{a_{n+1}}{a_n} \right| = \lim_{n \to \infty} \frac{n^2}{(n+1)^2} = 1$$

and hence $R = 1/\alpha = 1$.

Example 2

Find the radius of convergence of $\sum_0^\infty x^{2n}$.
    For this example, $a_n = 1$ for even n and $a_n = 0$ for odd n. Thus the limit in (2.21) is undefined. Nevertheless the series obviously has radius of convergence $R = 1$. This example shows that Equation (2.21) cannot always be used to determine R; a different, generally valid formula will be given in Section 5.5.

Exercises

1. Find the radii of convergence of the power series:

    (a) $\sum_0^\infty n^2 x^n$ (b) $\sum_0^\infty 2^{n+1} x^n$

    (c) $\sum_0^\infty \frac{x^n}{n!}$ (d) $\sum_0^\infty \frac{n^n}{n!} x^n$

2. Show that the series $\sum_0^\infty x^{2n}$ converges absolutely for $|x| < 1$ and diverges for $|x| > 1$.

3. The following power series all have radius of convergence $R = 1$. Discuss the convergence of the series at $x = \pm 1$.

    (a) $\sum_0^\infty x^n$ (b) $\sum_1^\infty \frac{x^n}{n}$ (c) $\sum_1^\infty \frac{x^n}{n^2}$

4. Discuss the convergence properties of a series of the form

$$\sum_{n=0}^\infty a_n(x - b)^n$$

    where b is a given real number.

5. Suppose $a_n = 0$ or 1 for each n (Exercise 2 is an example). What is the radius of convergence of $\sum_0^\infty a_n x^n$? (There are two cases.)

3 *Limits, Continuity, and Differentiability*

3.1 *Introduction*

The elementary study of limits of functions is strikingly similar to the study of limits of sequences. The main difference lies in replacing "$n \to \infty$" by "$x \to a$" in appropriate places. As in Chapter 1, we will begin with some simple numerical examples.

Example 1

It is "obvious" that $\lim_{x \to 3} x^2 = 9$. Determine a *positive* number δ such that:

$$|x^2 - 9| < \tfrac{1}{1000} \quad \text{provided } |x - 3| < \delta.$$

Solution   We write

$$|x^2 - 9| = |x - 3|\,|x + 3|.$$

It is easy to see that if $|x - 3| < 1$, for example, then $|x + 3| < 7$; this can be seen from Figure 3.1 and can be derived easily from the triangle inequality. See the lemma below.

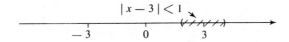

Figure 3.1

Therefore, we have

$$|x^2 - 9| = |x - 3||x + 3|$$

$$< 7|x - 3| \quad \text{if } |x - 3| < 1$$

$$< \frac{1}{1000} \quad \text{if } |x - 3| < \frac{1}{7000}.$$

Hence $\delta = 1/7000$ satisfies the requirement.

The following simple lemma is useful in calculations of "tolerances," δ.

Lemma   *If $|x - a| \leq h$, then*

$$\text{(a)} \quad |x - b| \leq |b - a| + h$$

and (3.1)

$$\text{(b)} \quad |x - b| \geq |b - a| - h.$$

The proof is immediate from the triangle inequality and is left to the reader (see Exercise 1). Both inequalities of (3.1) are also quite obvious from an examination of Figure 3.2.

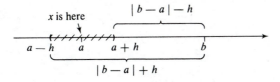

Figure 3.2

Example 2

Consider the formula

$$\lim_{x \to -1} \frac{x+2}{x-3} = -\frac{1}{4}.$$

Let $\varepsilon = 1/1000$. Find $\delta > 0$ such that

$$|x+1| = |x-(-1)| < \delta \quad \text{implies} \quad \left| \frac{x+2}{x-3} - \left(-\frac{1}{4}\right) \right| < \varepsilon.$$

Solution   First we see that

$$\left| \frac{x+2}{x-3} - \left(-\frac{1}{4}\right) \right| = \frac{5\,|x+1|}{4\,|x-3|}.$$

By (3.1)(b) we have

$$|x-3| \geq |3+1| - 1 = 3, \quad \text{provided } |x+1| \leq 1.$$

Therefore,

$$\left| \frac{x+2}{x-3} + \frac{1}{4} \right| = \frac{5\,|x+1|}{4\,|x-3|}$$

$$\leq \frac{5\,|x+1|}{12}, \quad \text{if } |x+1| \leq 1,$$

$$< \frac{1}{1000}, \quad \text{if } |x+1| < \frac{12}{5000} = 2.4 \times 10^{-3}.$$

Thus we can take $\delta = 2.4 \times 10^{-3}$.

In Examples 1 and 2 the values of the limits could be calculated simply by substitution—for example, $\lim_{x \to 3} x^2 = 3^2 = 9$. This is not the case, however, in many practical problems.

Example 3

Let

$$f(x) = \frac{1/x}{1 + 1/x} \quad (x \neq 0).$$

Determine $a = \lim_{x \to 0} f(x)$, and then determine $\delta > 0$ such that

$$|f(x) - a| < \frac{1}{4} \quad \text{if } |x| < \delta \ (x \neq 0).$$

Solution   Of course we can simplify $f(x)$ as follows:

$$f(x) = \frac{1/x}{1 + (1/x)} = \frac{1}{x + 1} \quad (x \neq 0),$$

and now it is clear that $\lim_{x \to 0} f(x) = 1$. To find δ, we have

$$|f(x) - 1| = \left| \frac{1}{x + 1} - 1 \right| = \frac{|x|}{|x + 1|} \quad (x \neq 0)$$

$$< \frac{|x|}{1/2}, \quad \text{if } |x| < \frac{1}{2},$$

$$< \frac{1}{4}, \quad \text{if } |x| < \frac{1}{8}.$$

Therefore, if we take $\delta = 1/8$ we get $|f(x) - 1| < 1/4$ as desired.

The above example may seem artificial, but we have to begin with very simple examples; less trivial ones will be encountered later. We consider finally an example in which the limit does not exist, so that it is impossible to determine δ.

Example 4

Let $f(x) = x/|x|$, and let L be a given real number. Show that no positive number δ exists such that

$$|f(x) - L| < 0.5 \quad \text{for all } x \text{ with } 0 < |x| < \delta.$$

(This shows that $\lim_{x \to 0} f(x) \neq L$ for any L.)

Solution Suppose that the inequality $|f(x) - L| < 0.5$ does hold for $0 < |x| < \delta$. Then, by the triangle inequality,

$$\left| f\left(\frac{\delta}{2}\right) - f\left(-\frac{\delta}{2}\right) \right| = \left| f\left(\frac{\delta}{2}\right) - L + L - f\left(-\frac{\delta}{2}\right) \right|$$

$$\leq \left| f\left(\frac{\delta}{2}\right) - L \right| + \left| f\left(-\frac{\delta}{2}\right) - L \right|$$

$$< 0.5 + 0.5 = 1.$$

Notice however (Figure 3.3) that $f(x) = 1$ if $x > 0$, and $f(x) = -1$ if $x < 0$.

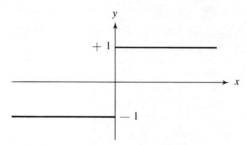

Figure 3.3 $f(x) = \dfrac{x}{|x|}$.

Therefore,

$$\left| f\left(\frac{\delta}{2}\right) - f\left(-\frac{\delta}{2}\right) \right| = 2.$$

This contradicts our supposition.

Exercises

1. Prove the lemma of this section.

2. In each of the following problems we have $\lim_{x \to a} f(x) = L$. Let $\varepsilon = 0.001$. For each case determine some positive number δ such that

$$|f(x) - L| < \varepsilon, \quad \text{provided } |x - a| < \delta \quad \text{and} \quad x \neq a.$$

(a) $\lim\limits_{x \to 0} (x + 1)^3 = 1$,

(b) $\lim\limits_{x \to 1/4} \dfrac{1}{x^2 + 2} = \dfrac{16}{33}$,

(c) $\lim\limits_{x \to 5} \left(\dfrac{1}{x} - 2x\right) = -\dfrac{49}{5}$,

(d) $\lim\limits_{x \to 2} \dfrac{x^2 + 4}{x + 2} = 2$,

(e) $\lim\limits_{x\to -1} \dfrac{1}{\sqrt{x^2+1}} = \dfrac{1}{\sqrt{2}}$,

(f) $\lim\limits_{x\to 8} x^{1/3} = 2$ (*Hint:* By algebra we have

$$x - 8 = (x^{1/3} - 2)(x^{2/3} + 2x^{1/3} + 4).),$$

(g) $\lim\limits_{x\to 1} \dfrac{x^2-1}{x-1} = 2$,   (h) $\lim\limits_{x\to 0} \dfrac{1-x^{-1}}{1+x^{-1}} = -1$.

3.   Find the following limits (if they exist) by inspection:

(a) $\lim\limits_{x\to 1} \dfrac{x^2-1}{x^3-1}$ (Factor!),   (b) $\lim\limits_{x\to 1} \dfrac{1-x}{1-\sqrt{x}}$,

(c) $\lim\limits_{x\to 0} \dfrac{x^3}{|x|}$,   (d) $\lim\limits_{x\to 0} e^{-1/x^2}$.

3.2   The Limit of a Function

In order for $\lim_{x\to a} f(x)$ to be meaningful, it is first necessary that the function $f(x)$ be defined for all values of x in some *neighborhood* of $x = a$ (except possibly the point $x = a$ itself). By a *neighborhood* of the point $x = a$ we mean an interval $(a - p, a + p)$ where $p > 0$. More precisely, the interval $(a - p, a + p)$ is called the *p-neighborhood* of $x = a$. Also, the interval $(a - p, a + p)$ with the point $x = a$ itself removed is called the *deleted p-neighborhood* of $x = a$. To say that $f(x)$ is defined on the deleted p-neighborhood of $x = a$ thus means that $f(x)$ is defined for all x satisfying $0 < |x - a| < p$, so that $f(a)$ itself need *not* be defined. We can now give the rigorous "ε-δ" definition of the limit of a function.

Definition   *Let $f(x)$ be defined in some deleted neighborhood of $x = a$. Then $\lim_{x\to a} f(x) = L$ means that for any given $\varepsilon > 0$ there is a corresponding $\delta > 0$ (which may depend on ε) such that*

$$|f(x) - L| < \varepsilon \quad \text{for all } x \text{ satisfying} \quad |x - a| < \delta, x \neq a. \qquad (3.2)$$

Let us emphasize that in the definition we do not assume that $f(a)$ is defined (but it may be!); therefore, in (3.2) the point $x = a$ is excluded from consideration. This is an important observation, particularly when it comes time to define derivatives in terms of limits. Also we do not suppose that ε (or δ) is "small." To do so would complicate the definition unnecessarily, and anyway we would then have to define "small." Logically speaking, if Condition (3.2) holds for all "small" $\varepsilon > 0$, it automatically holds for *all* $\varepsilon > 0$, and conversely.

Of course in practice it is usually the "small" values of ε which have to be especially taken into account.

If Condition (3.2) is satisfied, we say that the *limit of $f(x)$ as x approaches a exists* and equals L. (Naturally L is unique: see Theorem 1 below.)

Notice how similar (logically) this definition is to the definition of the limit of a sequence. We might say that the two definitions are the "same" except for certain necessary "adjustments." This similarity means also that the elementary theory of limits of functions closely resembles that of sequences.

We illustrate the procedure involved by a few examples.

Example 1

Show that $\lim_{x\to 3} x^2 = 9$.†

Solution   Let $\varepsilon > 0$. Then

$$|x^2 - 9| = |x + 3|\,|x - 3|$$
$$< 7\,|x - 3| \quad (\text{if } |x - 3| < 1)$$
$$< \varepsilon \qquad (\text{if also } |x - 3| < \varepsilon/7).$$

Let $\delta = \min(1, \varepsilon/7)$. Then we have shown that

$$|x^2 - 9| < \varepsilon, \quad \text{provided } |x - 3| < \delta.$$

Note that the simple choice $\delta = \varepsilon/7$ does not work for *every* $\varepsilon > 0$.

Example 1 is quite trivial, at least in the sense of evaluating $\lim_{x\to 3} x^2$. We will therefore give some less obvious examples. The reader would profit by trying to find the limits in question—for example, by examining the graph of $f(x)$—before reading the solutions.

Example 2

Find $\lim\limits_{x\to 0} \dfrac{x^2}{|x|}$, and prove it.

Solution   Notice that $|x^2/|x|| = |x|$ if $x \neq 0$, so it should be obvious that the limit is 0 (See Figure 3.4). In fact, if $\varepsilon > 0$ is given, we can take $\delta = \varepsilon$; then

$$\left|\frac{x^2}{|x|}\right| = |x| < \varepsilon \quad \text{if } |x| < \delta = \varepsilon \quad \text{and} \quad x \neq 0.$$

---

† You may have noticed that in many of our examples we have $\lim_{x\to a} f(x)$ equal to $f(a)$. As we shall see in Section 3.4, this situation is characteristic of *continuous* functions $f(x)$.

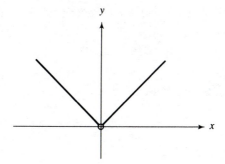

Figure 3.4 $f(x) = x^2/|x|.$†

Example 3

Define a function $g(x)$ for $-1 \leq x \leq 1$ as follows:

$$g(x) = \begin{cases} 1 & \text{if } x \text{ is any of the numbers } \pm 1, \pm \dfrac{1}{2}, \pm \dfrac{1}{3}, \ldots, \pm \dfrac{1}{n}, \ldots, \\ 0 & \text{otherwise.} \end{cases}$$

(See Figure 3.5). Find

     (a) $\lim\limits_{x \to 3/8} g(x)$;   (b) $\lim\limits_{x \to 1/4} g(x)$;   (c) $\lim\limits_{x \to 0} g(x)$.

Prove your answers.

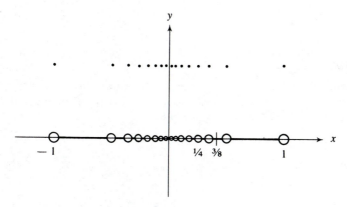

Figure 3.5 The function $g(x)$ of Example 3

---

† We use the convention, in sketching graphs, that a small circle on the graph indicates that the point encircled is not on the graph of $f(x)$.

Solution   (a) $\lim_{x\to 3/8} g(x) = 0$.   To prove this, let $\varepsilon > 0$ be given. Choose

$$\delta = \tfrac{3}{8} - \tfrac{1}{3} = \tfrac{1}{24}$$

(note that in this case δ does not depend on ε!). Then the interval $((3/8) - \delta,$ $(3/8) + \delta)$ contains none of the numbers $\pm 1, \pm 1/2, \pm 1/3, \ldots$ (why?), so that

$$|g(x)| = 0 < \varepsilon \quad \text{for all } x \text{ with } |x - \tfrac{3}{8}| < \delta.$$

This shows that $g(x) \to 0$ as $x \to 3/8$.

  (b) $\lim_{x\to 1/4} g(x) = 0$.   To see this, we take

$$\delta = \tfrac{1}{4} - \tfrac{1}{5} = \tfrac{1}{20}$$

(again independent of ε). Then for any given $\varepsilon > 0$,

$$|g(x)| = 0 < \varepsilon \quad \text{if } |x - \tfrac{1}{4}| < \delta \quad \text{and} \quad x \neq \tfrac{1}{4}.$$

(Note that $g(1/4) \neq 0$, but this does not affect the value of $\lim_{x\to 1/4} g(x)$.)

  (c) $\lim_{x\to 0} g(x)$ does not exist.   To show this, suppose we have $\lim_{x\to 0} g(x) = L$. Take $\varepsilon = 1/2$; then there must exist $\delta > 0$ such that

$$|g(x) - L| < \tfrac{1}{2} \quad \text{whenever } 0 < |x| < \delta.$$

But we can select two points x_1 and x_2 such that $0 < |x_i| < \delta$ ($i = 1, 2$), and $g(x_1) = 0$ and $g(x_2) = 1$. Therefore we have

$$|g(x_1) - L| = |L| < \tfrac{1}{2} \quad \text{and} \quad |g(x_2) - L| = |1 - L| < \tfrac{1}{2}.$$

Since both these inequalities cannot be satisfied by any real number L, we conclude that L does not exist.

  We proceed next to the most elementary theorems about limits of functions. These theorems, as well as their proofs, are closely analogous to theorems proved in Chapter 1 for limits of sequences.

Theorem 1   *If $\lim_{x\to a} f(x)$ exists, it is unique.*

Proof   Suppose L_1 and L_2 are both limits of $f(x)$ as $x \to a$. Then, given any $\varepsilon > 0$, there must exist a number $\delta > 0$ such that both

$$|f(x) - L_1| < \frac{\varepsilon}{2} \quad \text{and} \quad |f(x) - L_2| < \frac{\varepsilon}{2}$$

provided $0 < |x - a| < \delta$. But then

$$|L_1 - L_2| \leq |L_1 - f(x)| + |f(x) - L_2| < \frac{\varepsilon}{2} + \frac{\varepsilon}{2} = \varepsilon.$$

Since ε is arbitrary, we conclude that $L_1 = L_2$. ∎

Theorem 2   *Suppose that*

$$\lim_{x \to a} f(x) = L \quad and \quad \lim_{x \to a} g(x) = M.$$

Then

(i) $\lim\limits_{x \to a} [f(x) + g(x)] = L + M,$

(ii) $\lim\limits_{x \to a} f(x)g(x) = LM,$

(iii) $\lim\limits_{x \to a} \dfrac{f(x)}{g(x)} = \dfrac{L}{M} \quad if\ M \neq 0.$

We leave the proofs of (i) and (ii) as exercises. To prove (iii) we first establish the following lemma.

Lemma   *Suppose that* $\lim_{x \to a} g(x) = M \neq 0$. *Then there is some number* $\delta_0 > 0$ *such that*

$$|g(x)| > \tfrac{1}{2}|M| \quad for\ 0 < |x - a| < \delta_0.$$

Proof If $\delta_0 > 0$ is chosen so that

$$|g(x) - M| < \tfrac{1}{2}|M| \quad for\ 0 < |x - a| < \delta_0,$$

then we have

$$|g(x)| = |M - (M - g(x))|$$

$$\geq |M| - |M - g(x)|$$

$$> |M| - \tfrac{1}{2}|M| = \tfrac{1}{2}|M| \quad for\ 0 < |x - a| < \delta_0. \quad ∎$$

Proof of Theorem 2(iii) First consider the case $f(x) \equiv 1\dagger$. We have

$$\left| \frac{1}{g(x)} - \frac{1}{M} \right| = \frac{|g(x) - M|}{|M|\,|g(x)|}$$

$$< \frac{2\,|g(x) - M|}{M^2} \quad \text{if } 0 < |x - a| < \delta_0,$$

where $\delta_0 > 0$ is determined by the lemma. If $\varepsilon > 0$, let $\delta_1 > 0$ be chosen so that

$$|g(x) - M| < \frac{\varepsilon M^2}{2} \quad \text{for } 0 < |x - a| < \delta_1.$$

Finally, let $\delta = \min (\delta_0, \delta_1)$; then we have

$$\left| \frac{1}{g(x)} - \frac{1}{M} \right| < \varepsilon \quad \text{for } 0 < |x - a| < \delta.$$

This shows that $\lim_{x \to a} 1/g(x) = 1/M$.

To prove (iii) for an arbitrary numerator $f(x)$, we assume that part (ii) has already been proved. Then we know that

$$\lim_{x \to a} \frac{f(x)}{g(x)} = \lim_{x \to a} f(x) \cdot \frac{1}{g(x)} = \lim_{x \to a} f(x) \cdot \lim_{x \to a} \frac{1}{g(x)} = \frac{L}{M} . \quad \blacksquare$$

Theorem 3   *Given $f(x) \le g(x)$ for every x in some deleted neighborhood of $x = a$, suppose that*

$$\lim_{x \to a} f(x) = L \quad and \quad \lim_{x \to a} g(x) = M.$$

Then $L \le M$.

The proof, similar to the case of sequential limits, is left as an exercise. Note that if $f(x) < g(x)$ for every x in some neighborhood of $x = a$, we cannot necessarily conclude that $L < M$.

Exercises

1. Give a verbal description of the definition of $\lim_{x \to a} f(x) = L$, to accompany Figure 3.6.

$\dagger$ The notation $f(x) \equiv 1$ ("$f(x)$ is *identically equal* to 1") means that $f(x) = 1$ *for all x.*

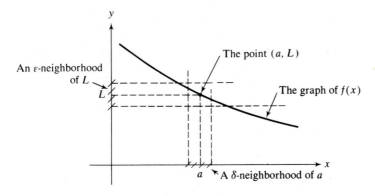

Figure 3.6 $\lim\limits_{x \to a} f(x) = L.$

2. Prove the following "obvious" limit formulas by the ε-δ method (cf. Exercise 1 of Section 3.1).

   (a) $\lim\limits_{x \to 0} (x + 1)^3 = 1,$

   (b) $\lim\limits_{x \to 2} \dfrac{x^2 + 4}{x + 2} = 2,$

   (c) $\lim\limits_{x \to -1} \dfrac{1}{\sqrt{x^2 + 1}} = \dfrac{1}{\sqrt{2}}\,.$

3. Make a reasonable sketch of the graph of $f(x) = \sqrt{|x|}$ on the interval $-1 \le x \le 1$. What is $\lim\limits_{x \to 0} f(x)$? Give the proof.

4. Show by the ε-δ method that $\lim\limits_{x \to 0} \dfrac{1}{\log |x|} = 0.$ (The inequalities needed may be confusing. Remember that $\log |x| < 0$ if $0 < |x| < 1$.)

5. Give an ε-δ proof of the fact that $\lim\limits_{x \to 0} \dfrac{1}{x}$ does not exist.

6. Prove Theorem 2(i).

7. (a) Define "*bounded*" for a function $f(x)$ defined on a given set A.
   (b) Prove the lemma: If $\lim\limits_{x \to a} f(x) = L$, then $f(x)$ is bounded on some deleted δ_0-neighborhood of a.

8. Prove Theorem 2(ii). (Use the lemma of Exercise 7.)

9. Suppose that $\lim\limits_{x \to a} f(x) = L.$
   (a) Prove that $\lim\limits_{x \to a} |f(x)| = |L|.$
   (b) Prove that if $L > 0$, then $\lim\limits_{x \to a} \sqrt{f(x)} = \sqrt{L}.$

10. Prove or disprove the following statements.

(a) $\lim_{x \to 2a} f(x) = 2 \lim_{x \to a} f(x),$ (b) $\lim_{x \to a} f(2x) = 2 \lim_{x \to a} f(x),$

(c) $\lim_{x \to 2a} f(x) = \lim_{x \to a} f(2x).$

11. Give the proof of Theorem 3.

3.3   Other Types of Limits

Sometimes one wishes to find the limit of a given function $f(x)$ as x approaches a from one side. For example, let $f(x) = e^{-1/x}$. Then, as we show below, $f(x) \to 0$ as $x \to 0$ with $x > 0$, whereas $\lim_{x \to 0} f(x)$ does not exist. See Figure 3.7.

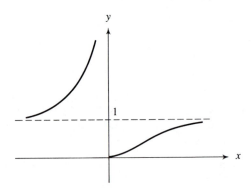

Figure 3.7 $f(x) = e^{-1/x}$.

The formal definition of one-sided limits is obtained by making the obvious adjustment to the basic definition of Section 3.2. Namely, we say that

$$\lim_{x \to a+} f(x) = L$$

if $f(x)$ is defined for $a < x < a + p$ (some $p > 0$) and the following condition holds:

for any given $\varepsilon > 0$ there is a number $\delta > 0$ such that

$$|f(x) - L| < \varepsilon \quad \text{whenever } a < x < a + \delta.$$

Example 1

$\lim_{x \to 0^+} e^{-1/x} = 0.$ To prove this, let $\varepsilon > 0$ be given. If $\varepsilon < 1$, then the required inequality

$$e^{-1/x} < \varepsilon$$

can be solved as follows:†

$$\log e^{-1/x} = -\frac{1}{x} < \log \varepsilon;$$

equivalently,

$$x \log \varepsilon > -1 \qquad \text{(since } x > 0 \text{ by assumption)};$$

that is,

$$x < -\frac{1}{\log \varepsilon} \quad \text{(since } \log \varepsilon < 0 \text{ for } \varepsilon < 1\text{)}.$$

Hence, if we take

$$\delta = -\frac{1}{\log \varepsilon} \quad \text{when} \quad \varepsilon < 1$$

and

$$\delta = \text{anything when } \varepsilon \geq 1,$$

then we have

$$e^{-1/x} < \varepsilon \quad \text{if } 0 < x < \delta.$$

(Why does this hold when $\varepsilon \geq 1$?)

Next we consider limits that involve infinity. There are several possibilities, since we may have $x \to +\infty$ or $-\infty$, or $f(x) \to +\infty$ or $-\infty$. In every case the formal definition is obtained by making a suitable adjustment in the basic limit definition. We will give one example, and ask the student to give other definitions himself (see the Exercises).

Definition   *Let $f(x)$ be a function defined for all $x > x_0$ for some x_0. We write*

$$\lim_{x \to +\infty} f(x) = L$$

if the following condition is satisfied:

given any $\varepsilon > 0$ there exists a real number M such that

$$|f(x) - L| < \varepsilon \quad \text{whenever } x > M.$$

---

† We use here the fact that $\log x$ is an increasing function of x.

Example 2

Find M such that

$$\left|\frac{x}{x+2} - 1\right| < \varepsilon \quad \text{for all } x > M.$$

$$\left(\text{This will prove that } \lim_{x \to +\infty} \frac{x}{x+2} = 1.\right)$$

Solution   This is easy.

$$\left|\frac{x}{x+2} - 1\right| = \frac{2}{x+2} < \frac{2}{x} < \varepsilon,$$

provided $x > \dfrac{2}{\varepsilon} = M$.

Example 3

$\lim_{x \to 0-} e^{-1/x} = +\infty$ (see Figure 3.7).
We can check this as follows:

$$\lim_{x \to 0-} e^{-1/x} = \lim_{t \to -\infty} e^{-t} \quad \text{(replacing } x \text{ by } 1/t)$$

$$= \lim_{u \to +\infty} e^{u} = +\infty.$$

The method of replacing x by $1/t$ in this example is easily justified by referring to the definitions. We give the following theorem as one example of this process.

Theorem 1   *Let $f(x)$ be defined on an interval $0 < x < a$. If either of the limits*

$$\lim_{x \to 0+} f(x) \quad \text{or} \quad \lim_{t \to +\infty} f\left(\frac{1}{t}\right)$$

exists, or if either limit is $+\infty$ or $-\infty$, then both limits have the same value.

Proof   Suppose for example that $\lim_{x \to 0+} f(x) = L$ $(L \neq \pm\infty)$. If $\varepsilon > 0$ is given, there exists $\delta > 0$ such that

$$|f(x) - L| < \varepsilon \quad \text{if } 0 < x < \delta.$$

Write $x = 1/t$; then

$$\left| f\left(\frac{1}{t}\right) - L \right| < \varepsilon \quad \text{if } t > \frac{1}{\delta}.$$

This shows that $\lim_{t \to +\infty} f(1/t) = L$. A similar argument applies if L is $+\infty$ or $-\infty$, or if the second limit is assumed to exist.   ∎

To write out a complete proof of the next theorem would be most tedious. By now you should be confident that you could prove any part of this theorem if necessary.

Theorem 2   *Let* $\lim f(x)$ *denote any one of the limits*

$$\lim_{x \to a+} f(x), \quad \lim_{x \to a-} f(x), \quad \lim_{x \to +\infty} f(x) \quad or \quad \lim_{x \to -\infty} f(x),$$

and assume $\lim f(x)$ *and* $\lim g(x)$ *exist and are finite. Then, under the usual hypotheses, all algebraic and order properties are preserved by taking this limit.*

Exercises

1. Find the following limits by inspection. Omit proofs. (Some limits may be infinite.)

   (a) $\lim\limits_{x \to 0} \log |x|,$

   (b) $\lim\limits_{x \to 0+} |\log x|,$

   (c) $\lim\limits_{x \to -\infty} \dfrac{x}{1 - 2x},$

   (d) $\lim\limits_{x \to \pi/2-} \tan x,$

   (e) $\lim\limits_{x \to +\infty} x^{-x},$

   (f) $\lim\limits_{x \to +\infty} \dfrac{\sin x}{x},$

   (g) $\lim\limits_{x \to +\infty} \arctan x,$

   (h) $\lim\limits_{x \to 2-} \dfrac{1}{x - 2}.$

2. Let $f(x) = \dfrac{e^{1/x}}{1 + e^{1/x}} \ (x \neq 0)$. Find

   (a) $\lim\limits_{x \to 0+} f(x),$

   (b) $\lim\limits_{x \to 0-} f(x),$

   (c) $\lim\limits_{x \to +\infty} f(x),$

   (d) $\lim\limits_{x \to -\infty} f(x).$

   (Use Examples 1 and 3 of the text.) Sketch the graph of $f(x)$.

3. Write down the formal definition for each of the following limits. L and a always denote finite numbers.

    (a) $\lim\limits_{x \to a-} f(x) = L,$

    (b) $\lim\limits_{x \to a} f(x) = +\infty,$

    (c) $\lim\limits_{x \to a} f(x) = -\infty,$

    (d) $\lim\limits_{x \to a+} f(x) = +\infty,$

    (e) $\lim\limits_{x \to +\infty} f(x) = +\infty.$

4. Show that $\lim_{x \to +\infty} f(x) = \lim_{t \to -\infty} f(-t)$ if either of these exists.

5. Prove that if $\lim_{x \to a} f(x) = +\infty$ and $\lim_{x \to a} g(x) = L$ (finite), then

    $$\lim_{x \to a} [f(x) + g(x)] = +\infty.$$

    (The following proof, though short, is unacceptable:

    $$\lim_{x \to a} [f(x) + g(x)] = \lim_{x \to a} f(x) + \lim_{x \to a} g(x) = +\infty + L = +\infty.)$$

6. Give the complete proof of the statement: If $f(x) \le g(x)$ for all x, then

    $$\lim_{x \to +\infty} f(x) \le \lim_{x \to +\infty} g(x),$$

    assuming these limits exist and are finite. Do the limits have to be finite?

7. What can be concluded if one knows that $\lim_{x \to a+} f(x) = \lim_{x \to a-} f(x)$?

8. Is it true or false that in every case

    $$\lim_{x \to +\infty} f(x) = \lim_{n \to +\infty} f(n)?$$

    (This may be controversial! Try to clarify the problem.)

3.4   *Continuity*

From early times man has pondered the question of the continuous. Mathematicians still refer to the real number system as the "continuum"; physicists talk about the "space-time continuum." The question of the continuous versus the discrete nature of *matter* was hotly argued by scientists and philosophers for centuries. These arguments carried over into pure mathematics, and led to such confusing notions as "infinitesimal," "instantaneous," and so on. One of the principal causes of confusion was the failure to separate the mathematical abstractions from physical reality. A completely satisfactory definition and theory of continuity was finally developed by nineteenth-century mathematicians, particularly K. Weierstrass (1815–1897) and J. Dedekind (1831–1916).

The first step in reaching the definition seems to have been the realization that continuity should refer always to *functions*, rather than to vaguely defined objects such as curves or surfaces. (In fact, nowadays, curves and surfaces are themselves defined in terms of continuous functions.) Most people have an intuitive idea of what a continuous graph is—a graph is continuous if it is "unbroken." As in Figure 3.8, this property seems to refer to the whole graph.

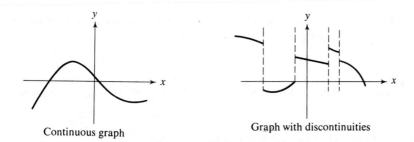

Continuous graph   Graph with discontinuities

Figure 3.8

The next step in understanding continuity is the realization that if a graph is broken *at a single point*, it is discontinuous (at that point, at least!), but otherwise it is continuous.†

Continuity at a given point $x = a$, however, can immediately be defined in terms of the limit as $x \to a$:

---

† It is interesting to note that the Russian word for continuous is nye'prerivni, which means, literally, un'broken.

Definition 1   *Let $f(x)$ be a function defined in some neighborhood of $x = a$. Then $f(x)$ is said to be **continuous at** $x = a$ provided that*

$$\lim_{x \to a} f(x) = f(a).$$

*The function $f(x)$ is said to be **continuous on a given interval** (α, β) if it is continuous at each point of this interval.*

By directly incorporating the ε-δ definition of the limit of a function, we see that the above definition of continuity at a point can be expressed in the following way (see condition (3.2)):

Equivalent Definition 1   *Let $f(x)$ be a function defined in some neighborhood of $x = a$. Then $f(x)$ is said to be **continuous at** $x = a$ provided that*

for any given $\varepsilon > 0$ there is a corresponding $\delta > 0$ (which may depend on ε) such that $|f(x) - f(a)| < \varepsilon$ for all x satisfying $|x - a| < \delta$.

Most functions of elementary calculus are continuous. For example, we can immediately prove the following theorem.

Theorem 1   *Every polynomial function $p(x)$ is continuous at every point.*

Proof   First, it is completely trivial to prove that

$$\lim_{x \to a} x = a.$$

Now let $p(x) = c_0 x^n + c_1 x^{n-1} + \cdots + c_n$ be a given polynomial. Then by the basic theorem (Section 3.3) about sums and products, we conclude that

$$\lim_{x \to a} p(x) = \lim_{x \to a} (c_0 x^n + c_1 x^{n-1} + \cdots + c_n)$$

$$= c_0 a^n + c_1 a^{n-1} + \cdots + c_n = p(a).$$

This proves that $p(x)$ is continuous at $x = a$.   ∎

Example 1

Let $f(x) = |x|$ (Figure 3.9). Since $\lim_{x \to a} |x| = |a|$ for any number a (even including $a = 0$), it follows that $f(x)$ is continuous everywhere.†

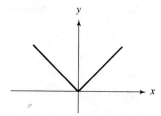

Figure 3.9 $f(x) = |x|$.

A function which is not continuous at a point $x = a$ is said to have a *discontinuity* at that point. There are several possible kinds of discontinuities, and the next several examples illustrate progressively "worse" cases.

Example 2

(A "trivial," or "removable," discontinuity.) Let

$$f(x) = x^{-1}\left(1 + \frac{1}{x}\right)^{-1} \quad (x \neq 0, -1),$$

as in Figure 3.10. Since $f(0)$ is not defined, the definition of continuity at $x = 0$ cannot be satisfied. But we have $\lim_{x \to 0} f(x) = 1$. We could therefore "extend"

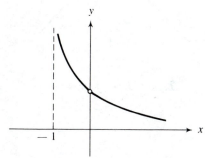

Figure 3.10 $f(x) = x^{-1}\left(1 + \frac{1}{x}\right)^{-1}$.

---

† However, note that $f(x) = |x|$ is not *smooth* at $x = 0$; that is, it does not have a derivative at $x = 0$. See Section 3.5.

the definition of $f(x)$ by defining $f(0) = 1$. The extended function $f(x)$ *is* continuous at $x = 0$ (but not at $x = -1$).

Definition 2   *A function $f(x)$ defined for all values of x in some deleted neighborhood of $x = a$ is said to have a **trivial** or **removable discontinuity** at $x = a$ if*

$$\lim_{x \to a} f(x) = L$$

exists, but $L \neq f(a)$. In this case the definition of $f(x)$ can be modified to obtain a function continuous at $x = a$, by (re)defining $f(a) = L$.

Example 3

Let $f(x) = \dfrac{\sin x}{x}$ $(x \neq 0)$. It is proved in elementary calculus texts that

$$\lim_{x \to 0} \frac{\sin x}{x} = 1.$$

Hence, $f(x)$ has a removable discontinuity at $x = 0$ (see Figure 3.11).

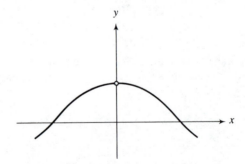

Figure 3.11 $f(x) = \dfrac{\sin x}{x}$.

Example 4

Let $f(x) = \dfrac{x}{|x|}$ $(x \neq 0)$. It is easy to see (Figure 3.12) that

$$\lim_{x \to 0+} f(x) = +1 \quad \text{and} \quad \lim_{x \to 0-} f(x) = -1.$$

This is an example of a function having a "jump discontinuity" at $x = 0$.

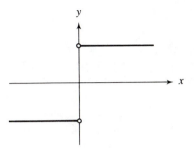

Figure 3.12 $f(x) = \dfrac{x}{|x|}$.

Definition 3 *A function $f(x)$ defined in a deleted neighborhood of $x = a$ is said to have a **jump discontinuity** at $x = a$ if $\lim_{x \to a+} f(x)$ and $\lim_{x \to a-} f(x)$ exist (finitely), but are not equal. The value*

$$J_a(f) = \lim_{x \to a+} f(x) - \lim_{x \to a-} f(x)$$

is called the **jump of f at a.**

Another example of a jump discontinuity is provided by the function (see Exercise 2, Section 3.3)

$$f(x) = \frac{e^{1/x}}{1 + e^{1/x}}.$$

Example 5

Let $f(x) = \sin 1/x$ $(x \neq 0)$. This funtion has an *oscillating discontinuity* at $x = 0$. In this case the oscillations are *bounded*.

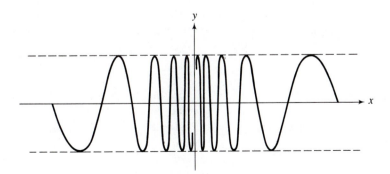

Figure 3.13 $f(x) = \sin \dfrac{1}{x}$.

Example 6

The functions shown in Figure 3.14 illustrate various kinds of *infinite discontinuities* at $x = 0$. The reader can provide formal definitions of these sorts of discontinuities if desired.

In all of the above examples, the functions $f(x)$ have only finitely many points of discontinuity. We next give two much more complicated examples, in which there are infinitely many points of discontinuity in a finite interval.

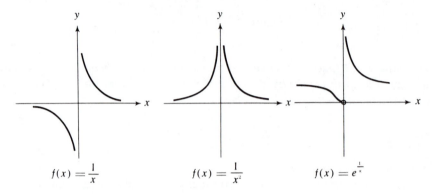

$$f(x) = \frac{1}{x} \qquad\qquad f(x) = \frac{1}{x^2} \qquad\qquad f(x) = e^{\frac{1}{x}}$$

Figure 3.14

Example 7

Define a function $f(x)$ as follows:

$$f(x) = \begin{cases} 1 & \text{if } x \text{ is rational,} \\ 0 & \text{if } x \text{ is irrational.} \end{cases}$$

Since any interval $(a - \varepsilon, a + \varepsilon)$, $\varepsilon > 0$, contains points at which $f(x) = 0$ and points at which $f(x) = 1$, it is clear that $\lim_{x \to a} f(x)$ does not exist. Hence $f(x)$ is not continuous at any point $x = a$.

Example 8

(Dirichlet's function).   Define a function $f(x)$ on $0 < x < 1$ by setting

$$f(x) = \begin{cases} \dfrac{1}{q} & \text{if } x = \dfrac{p}{q} \text{ in lowest terms,} \\ 0 & \text{if } x \text{ is irrational.} \end{cases}$$

Some examples of values of $f(x)$ are:

$$f(\tfrac{3}{5}) = \tfrac{1}{5}, \quad f(\tfrac{99}{100}) = \tfrac{1}{100}, \quad f\left(\frac{p\sqrt{2}}{q}\right) = 0,$$

and so on. It is easy to see that the total number of points x, $0 < x < 1$, at which $f(x) \geq 1/q$ is at most $(1/2)\,q(q-1)$. See Figure 3.15. Let us show that for

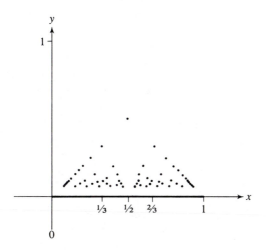

Figure 3.15 Dirichlet's function.

the Dirichlet function we have

$$\lim_{x \to a} f(x) = 0 \qquad\qquad (3.3)$$

for every a in $(0, 1)$. Since $f(a) = 0$ if and only if a is irrational, this will prove that *Dirichlet's function is continuous at every irrational number and discontinuous at every rational number.*

To prove (3.3), let $\varepsilon > 0$ be given. Choose a positive integer q such that $1/q < \varepsilon$. Then we can find some positive number δ such that the interval $(a - \delta, a + \delta)$ contains none of the finitely many points x at which $f(x) \geq 1/q$, except possibly the point a itself. (Can you show how to determine δ?) This means that

$$|f(x)| = f(x) < \frac{1}{q} < \varepsilon$$

for all x with $|x - a| < \delta$, $x \neq a$, since for such values of x, either $f(x) = 0$ or $f(x) = 1/r$ with $r > q$. This completes the proof of (3.3).

Let us return to the study of functions that *are* continuous. Since continuity is defined in terms of limits, and since sums, products, and quotients are preserved by the limit operation, we obtain the next result.

Theorem 2   *Suppose that $f(x)$ and $g(x)$ are continuous at $x = a$. Then the functions*

(i) $f(x) + g(x),$

(ii) $f(x)g(x),$

(iii) $f(x)/g(x)$ *(if $g(a) \neq 0$)*

are also continuous at $x = a$.

Proof To prove (i), for example, we have

$$\lim_{x \to a} [f(x) + g(x)] = \lim_{x \to a} f(x) + \lim_{x \to a} g(x)$$
$$= f(a) + g(a),$$

by hypothesis. This proves that $f(x) + g(x)$ is continuous at $x = a$. The proofs of (ii) and (iii) are similar. ∎

Next we consider the case of a *composite* function $f(g(x))$.

Theorem 3 *Let $g(x)$ be continuous at $x = a$, and $f(t)$ continuous at $t = g(a)$. Then the composite function $f(g(x))$ is continuous at $x = a$.*

The function $f(g(x))$ is called the *composite function of f and g*; Theorem 3 says that the composite of two continuous functions is again a continuous function.

Proof We have to show that

$$\lim_{x \to a} f(g(x)) = f(g(a)). \tag{3.4}$$

Let $\varepsilon > 0$ be given. Since $f(t)$ is continuous at $t = g(a)$, there exists $\delta_1 > 0$ such that

$$|f(t) - f(g(a))| < \varepsilon \quad \text{if } |t - g(a)| < \delta_1. \tag{3.5}$$

But since $g(x)$ is continuous at $x = a$, there exists $\delta_2 > 0$ such that if $|x - a| < \delta_2$, then

$$|g(x) - g(a)| < \delta_1.$$

Hence by (3.5) with $t = g(x)$,

$$|f(g(x)) - f(g(a))| < \varepsilon \quad \text{if } |x - a| < \delta_2.$$

This proves (3.4). ∎

Example 9

Let $f(x) = \sqrt{|(x - 1)(x - 2)|}$ (see Figure 3.16). We know that

   (a) $(x - 1)(x - 2)$ is continuous for all x (Theorem 1);

   (b) $|t|$ is continuous for all t (Example 1); and

   (c) $\sqrt{u}$ is continuous for all $u \geq 0$ (Theorem 4 below).†

   Hence, applying Theorem 3 twice, we conclude that $f(x)$ is continuous for all x.

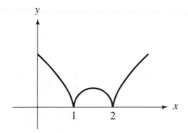

Figure 3.16 $f(x) = \sqrt{|(x - 1)(x - 2)|}$.

Theorem 4 *The following functions are continuous:*

   (i) $\sin x$ *and* $\cos x$, *for all* x;

   (ii) e^x, *for all* x;

   (iii) $\log x$, *for all* $x > 0$;

   (iv) x^p *(where p is any real number), for all* $x > 0$.

---

   † Actually, continuity of a function such as $\sqrt{u}$, which is defined only for $u \geq 0$, has not yet been defined at $u = 0$. However, the reader can easily provide the necessary definition, using one-sided limits (see Exercise 6). See also Section 4.5.

The proof of Theorem 4 will not be given here. Later, in Chapter 5, we will define the exponential, logarithmic, and trigonometric functions carefully and prove that they are continuous. Meanwhile, although we will not hesitate to use these functions in examples and exercises, we will avoid using them in the logical development of the theory, thereby avoiding the danger of logical "circularity."

Exercises

1. Each of the following expressions defines a function $f(x)$ for all $x \in \mathbb{R}$ with finitely many exceptions. Determine all these exceptional points and investigate the type of discontinuity that $f(x)$ possesses at each exceptional point. For removable discontinuities, calculate $\lim_{x \to a} f(x)$. For jumps, calculate $J_a(f)$.

(a) $\dfrac{1}{|x| - 1}$,

(b) $\dfrac{x - 1}{|x| - 1}$,

(c) $\dfrac{x - 1}{|x - 1|}$,

(d) $\dfrac{1}{\sin x}$,

(e) $\dfrac{\sin x}{\sqrt{|x|}}$,

(f) $x \sin \dfrac{1}{x}$,

(g) $[10x]$ (Note: $[10x]$ is the greatest integer $\leq 10x$.)

(h) $\log \left| \dfrac{1}{x} \right|$,

(i) $\dfrac{1}{\log|x|}$,

(j) $\dfrac{1}{x} \sin \log|x|$.

2. Let $\Psi(T)$ denote the amount of heat which must be applied to raise a given block of ice at $0°F$ to a temperature of $T°F$. For which values of T is $\Psi(T)$ discontinuous, and what is the physical significance of the discontinuity?

3. Dirichlet's function (Example 8) is discontinuous at $x = 1/2$. What type of discontinuity is this?

4. Sketch carefully the graph of $f(x) = x^{2/3}$ near the origin. Does this function appear to be continuous at the origin? Is it?

5. Let $0 < x_1 < x_2 < \cdots < x_n < 1$ be finitely many points. Define functions

$$a_k(x) = \begin{cases} 0 & \text{if } x < x_k, \\ 1 & \text{if } x \geq x_k \end{cases}$$

(for $k = 1, 2, \ldots, n$). Sketch the graph of

$$f(x) = \frac{1}{n} \sum_{k=1}^{n} a_k(x)$$

and discuss its continuity.

6. If $f(x)$ is a function defined for $a \leq x \leq a + \delta$ (some $\delta > 0$), we say that $f(x)$ is *right-continuous* at $x = a$ if

$$\lim_{x \to a+} f(x) = f(a).$$

Discuss in terms of right continuity

(a) $[x]$, the greatest integer $\leq x$,
(b) the function of Exercise 5.

Define *left continuity* and discuss the same examples.

7. Give an example of a function $f(x)$ defined for all real x, but continuous only at $x = 0$.

8. Give an example of a function $f(x)$ which is continuous nowhere, but with $(f(x))^2$ continuous everywhere.

9. Define $f(x)$ by the formula

$$f(x) = \lim_{n \to \infty} \frac{x^n}{1 + x^n}.$$

Calculate $f(x)$ for all possible x and discuss its continuity.

10. Let $f(x)$ be defined and continuous for $a < x < b$. Suppose that $f(x) = 0$ for every rational number $x \in (a, b)$. Show that $f(x) = 0$ for all $x \in (a, b)$.

3.5 *Differentiability*

Historically the notion of the derivative of a function preceded by about two centuries the notion of continuity of a function. This observation is reflected in the relative ease with which students learn about derivatives, in comparison to the difficulty they encounter in understanding continuity.

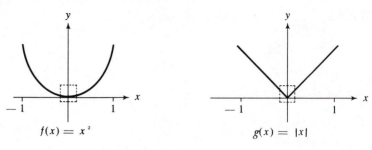

Figure 3.17

In this section we review the definition of derivative and prove that *a function f(x) is differentiable at a given point x = a if and only if the graph of f(x) "looks arbitrarily like a straight line" in sufficiently small neighborhoods of x = a.* This very important characterization of differentiability is seldom stressed in calculus books.

To illustrate this idea of differentiability, we consider the following two simple examples (Figure 3.17):

(a) $f(x) = x^2$,

(b) $g(x) = |x|$.

We know that both $f(x)$ and $g(x)$ are continuous at $x = 0$. To see how the graphs look in a "small" neighborhood of $x = 0$, let us magnify each graph ten times; see Figure 3.18. Note that "near" $x = 0$, the graph of $f(x)$ is

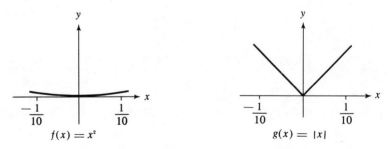

Figure 3.18

approximately a straight line, but $g(x)$ has the same sharp corner at $x = 0$, no matter how close one is to $x = 0$. In fact $f(x) = x^2$ is differentiable at $x = 0$, but $g(x) = |x|$ is not. Differentiable functions are sometimes said to have "smooth" graphs. We now transpose the foregoing description into mathematical terms.

Definition 1   *Let $f(x)$ be defined in a neighborhood of $x = a$. The **derivative of** $f(x)$ at $x = a$ is defined as*

$$f'(a) = \lim_{x \to a} \frac{f(x) - f(a)}{x - a},$$

*provided this limit exists. If the limit does exist, we say that $f(x)$ is **differentiable** at $x = a$.*

Theorem 1   *If $f(x)$ is differentiable at $x = a$, then $f(x)$ is continuous at $x = a$.*

Proof Since $\lim\limits_{x \to a} \dfrac{f(x) - f(a)}{x - a}$ exists, the function $\dfrac{f(x) - f(a)}{x - a}$ must be bounded near $x = a$:

$$\left| \frac{f(x) - f(a)}{x - a} \right| \leq M = \text{const} \quad (0 < |x - a| < \delta_0).$$

We can assume that $M \neq 0$.

If $\varepsilon > 0$ is given, let $\delta_1 = \varepsilon/M$. Then if $0 < |x - a| < \delta = \min(\delta_0, \delta_1)$, we have

$$|f(x) - f(a)| \leq M |x - a| < \varepsilon.$$

This proves that $\lim_{x \to a} f(x) = f(a)$. ∎

Definition 2   *Let $f(x)$ be defined in a neighborhood of $x = 0$. We say that $f(x)$ is of **smaller order than** x as $x \to 0$ provided that:*

for any given $\varepsilon > 0$ there exists $\delta > 0$ such that

$$|f(x)| \leq \varepsilon |x| \quad \text{whenever} \quad |x| < \delta. \tag{3.6}$$

If $f(x)$ is of smaller order than x as $x \to 0$, we write

$$f(x) = o(x) \quad \text{as } x \to 0.$$

This is read "$f(x)$ is little-oh of x as x approaches zero." (Compare Section 1.10, where a similar notation was introduced for sequences.)

Example 1

Let $f(x) = x^2$. Then

$$|f(x)| = |x^2| = |x| \cdot |x| \leq \varepsilon |x|$$

provided that $|x| < \varepsilon$. Hence, we may choose $\delta = \varepsilon$ in order to satisfy (3.6), so we conclude that x^2 is of smaller order than x as $x \to 0$.

Example 2

Let $f(x) = |x|$, and suppose $0 < \varepsilon < 1$. Then the inequality

$$|f(x)| = |x| \le \varepsilon |x|$$

cannot hold except for $x = 0$. This means that $|x|$ is not of smaller order than x as $x \to 0$.

Graphically, Condition (3.6) simply means that the curve $y = f(x)$ lies between the lines $y = \pm \varepsilon x$ when $|x| < \delta$ (see Figure 3.19). If $f(x)$ is a function of smaller order than x as $x \to 0$, we say that the curve $y = f(x)$ is *tangent to the x-axis at the origin.* This leads to the definition of tangent in general, as follows.

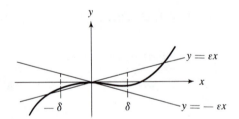

Figure 3.19 $|f(x)| < \varepsilon |x|$ for $|x| < \delta$.

Definition 3   *Let $f(x)$ be defined in a neighborhood of $x = a$. The curve $y = f(x)$ is said to be **tangent to the straight line***

$$y = m(x - a) + b \tag{3.7}$$

at $x = a$, provided that $f(x) - m(x - a) - b$ is of smaller order than $x - a$ as $x \to a$. In other words, the curve $y = f(x)$ is tangent to the line (3.7) if, given $\varepsilon > 0$, there exists $\delta > 0$ such that

$$|f(x) - m(x - a) - b| \le \varepsilon |x - a| \quad \text{when} \quad |x - a| < \delta. \tag{3.8}$$

Thus we see that the definition of tangent requires an ε-δ statement.†

---

† Definition 3 can be generalized in a very simple and obvious way to define tangent planes to surfaces $z = f(x, y)$, and even to more general situations in n dimensions.

The inequality (3.8) means graphically that the curve $y = f(x)$ lies between the lines $y - b = m(x - a) \pm \varepsilon(x - a)$ for $|x - a| < \delta$ (see Figure 3.20). Since ε can be arbitrary, the intuitive interpretation of the tangent line $y - b = m(x - a)$ is clear from this figure.

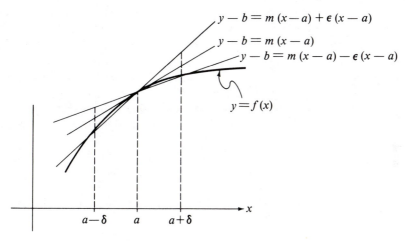

Figure 3.20 Condition (3.8).

Theorem 2 Let $f(x)$ *be defined in a neighborhood of* $x = a$. *Then the curve* $y = f(x)$ *is tangent to the line* $y = m(x - a) + b$ *at* $x = a$ *if and only if* $f(x)$ *is differentiable at* $x = a$ *and* $f(a) = b$ *and* $f'(a) = m$.

Proof First assume $y = f(x)$ is tangent to $y = m(x - a) + b$ at $x = a$. If we put $x = a$ in (3.8), we get

$$|f(a) - b| \le 0,$$

so that $f(a) = b$. Using the inequality (3.8) once again, we now have, for each $\varepsilon > 0$, a $\delta > 0$ such that

$$\frac{|f(x) - m(x - a) - f(a)|}{|x - a|} = \left| \frac{f(x) - f(a)}{x - a} - m \right| < \varepsilon$$

when $0 < |x - a| < \delta$. But this means precisely that

$$\lim_{x \to a} \frac{f(x) - f(a)}{x - a} = m.$$

Thus $f(x)$ is differentiable at $x = a$ and $f'(a) = m$.

The converse is proved simply by tracing backward the steps of the above proof. ∎

Example 3

The Lions Gate bridge at Vancouver, B.C. (Plate 1) provides an interesting illustration of nondifferentiability. The bridge is famous partly because of a sudden "bump" in the roadway at the center of the span. In fact, the curve of the roadway is nondifferentiable at this point. The sudden change in the slope of the roadway is quite evident to anyone driving over the bridge. (You may wish to ponder to what extent this description is a mathematical abstraction of the actual physical condition of the bridge.) By the way, the nondifferentiability of the curve of the supporting cable also shows up in the photo.

Let us now summarize the relationship between: (a) the existence of the *derivative* of $f(x)$ at $x = a$; (b) *differentiability* of $f(x)$ at $x = a$; (c) *linear approximability* of $f(x)$ near $x = a$; and (d) the existence of a *tangent line* to the curve $y = f(x)$ at $x = a$. In fact these concepts are all equivalent, and they can all be expressed by the single condition

$$f(x) - f(a) = m(x - a) + o(x - a) \quad \text{as } x \to a, \text{ with } m = f'(a).$$

Another related concept is that of the "differential," which we now define.

Definition 4   Let $f(x)$ be differentiable at $x = a$. The **differential** df **of** $f(x)$ **at** $x = a$ is then defined to be

$$df = df(a, dx) = f'(a)\, dx. \tag{3.9}$$

In Equation (3.9), the symbol dx represents an *arbitrary real number*. In practice, however, one tends to think of dx as being an "infinitesimal" (whatever that is!). This is because one treats (3.9) as an "approximate" formula, valid (approximately!) for "small" values of dx.

Writing

$$\Delta f(a, dx) = f(a + dx) - f(a)$$

we now see that this idea can be expressed rigorously as: let $f(x)$ be differentiable at $x = a$; then

$$\Delta f = df + o(dx) \qquad \text{as } dx \to 0.$$

In other words, for differentiable functions $f(x)$ the differential df is simply the linear approximation to Δf valid near $x = a$.

Suppose next that $f(x)$ is defined on an open interval (a, b) and that $f'(x)$ exists for each x in (a, b). Then $f'(x)$ is a function on (a, b), and we can consider whether $f'(x)$ is continuous or differentiable, and so on.

Plate 1 (Courtesy of The Vancouver Public Library, Historic Photograph Section.)

Definition 5   *Let $f(x)$ be defined on an open interval (a, b). We say that $f(x)$ is **continuously differentiable** on (a, b) if $f'(x)$ is defined and continuous on (a, b). More generally, we say that $f(x)$ is k **times continuously differentiable** on (a, b) if the kth derivative*

$$f^{(k)}(x) = \frac{d^k f(x)}{dx^k}$$

exists and is continuous on (a, b). $\left(\text{The kth derivative of } f(x) \text{ can be defined by}\right.$

recursion: $f^{(1)}(x) = f'(x)$; $f^{(k+1)}(x) = \dfrac{d}{dx} f^{(k)}(x)$ for $k = 1, 2, 3, \ldots$ $\left.\right)$

Example 4

The function $f(x)$ defined by

$$f(x) = \begin{cases} x^2 & \text{if } x \geq 0, \\ 0 & \text{if } x < 0 \end{cases}$$

is continuously differentiable, but not twice continuously differentiable, on $(-\infty, \infty)$.

To see this, we notice that

$$f'(x) = \begin{cases} 2x & \text{if } x \geq 0, \\ 0 & \text{if } x < 0, \end{cases}$$

so that $f'(x)$ is a continuous function, but not a differentiable function. See Figure 3.21.

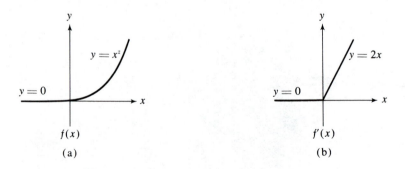

(a)   (b)

Figure 3.21

Example 5

As an application of continuous differentiability we consider the problem of designing highway curves. Let the curve $y = f(x)$ represent the center line of a (level) highway. It is certain that $f(x)$ must be continuous, and the most elementary dynamical considerations show that $f(x)$ must also be differentiable— see Figure 3.22(a). A little more reflection on the dynamics of an automobile leads to the conclusion that the curve $f(x)$ should be at least *twice differentiable*.

Highway engineers construct twice differentiable curves by fitting together cubic curves $(f(x) = ax^3 + bx^2 + cx + d)$. They may be unaware of the possibility of designing *infinitely differentiable* highways by utilizing the principle of the following example.

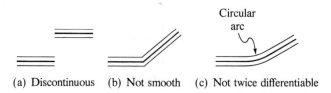

(a) Discontinuous   (b) Not smooth   (c) Not twice differentiable

Figure 3.22 Inadequate highway design.

Example 6

Let $f(x)$ be defined by

$$f(x) = \begin{cases} e^{-1/x}, & \text{if } x > 0, \\ 0, & \text{if } x \leq 0. \end{cases}$$

Then $f(x)$ is infinitely differentiable for all x; that is, $f^{(k)}(x)$ exists for every positive integer k.

To see this, note first that, from calculus, $f^{(k)}(x)$ exists for any k if $x \neq 0$. We can show that

$$f^{(k)}(0) = 0 \quad (k = 1, 2, 3, \ldots),$$

and this will prove the assertion. For $k = 1$ we have

$$f'(0) = \lim_{x \to 0} \frac{f(x) - f(0)}{x} = \lim_{x \to 0} \frac{f(x)}{x}.$$

Since $f(x) \equiv 0$ for $x < 0$, obviously $\lim_{x \to 0-} f(x)/x = 0$. For $x \to 0+$ we have

$$\lim_{x \to 0+} \frac{f(x)}{x} = \lim_{x \to 0+} \frac{1}{x} e^{-1/x} = \lim_{t \to +\infty} te^{-t} = 0,$$

by a well-known property of exponentials. Hence $f'(0) = 0$. Similarly

$$f''(0) = \lim_{x \to 0} \frac{f'(x) - f'(0)}{x} = \lim_{x \to 0} \frac{f'(x)}{x}.$$

The limit from the left is again obviously zero; since $f'(x) = x^{-2}e^{-1/x}$ for $x > 0$, we find that

$$\lim_{x \to 0+} \frac{f'(x)}{x} = \lim_{x \to 0+} x^{-3}e^{-1/x} = \lim_{t \to +\infty} t^3 e^{-t} = 0.$$

Therefore, $f''(0) = 0$. It is fairly easy to extend this argument, and prove by induction that $f^{(k)}(0) = 0$ for all k.

Exercises

1. Determine which of the following functions are of smaller order than x as $x \to 0$. Give reasons.

    (a) $x^{3/2}$,

    (b) $|x|^{1/2}$,

    (c) $\sin^2 x$,

    (d) $\dfrac{x^2}{x+1}$.

2. Each of the following functions is continuous everywhere. Determine all points at which each function is nondifferentiable. (A sketch of the graph will be quite revealing.)

    (a) $\dfrac{1}{|x| + 1}$;

    (b) $e^{|x|}$;

    (c) $|\sin x|$;

    (d) $|x| + |x - 2|$.

3. Give an example of a function which is k, but not $k + 1$, times continuously differentiable.

4. The *right-hand derivative* of $f(x)$ at $x = a$ is defined to be

    $$f'_+(a) = \lim_{x \to a+} \frac{f(x) - f(a)}{x - a}.$$

    Find the right- and left-hand derivatives of each of the functions (a)–(d) of Exercise 2, at each of its points of nondifferentiability.

5. Explain why the curve of Figure 3.22(c) is not suitable for a high-speed highway. (How would you steer an automobile through such a curve?)

6. Complete the proof by induction, that $f^{(k)}(0) = 0$ for all k, where $f(x)$ is the function of Example 6.

7. Let $f(x)$ denote the infinitely differentiable function of Example 6.
   (a) Sketch the graph of $f(x)$.
   (b) Sketch the graph of $g(x) = f(1 + x) \cdot f(1 - x)$.
   (c) Show that $g(x)$ is infinitely differentiable for all x.
   (d) The *support* of a function $g(x)$ is defined as the set of points x where $g(x) \neq 0$. What is the support of the above function?

*8. Prove that the function $f(x)$ defined by

$$f(x) = x^2 \sin \frac{1}{x} \quad (x \neq 0)$$

$$f(0) = 0$$

is differentiable, but not continuously differentiable, at $x = 0$. Also, sketch the graph of $f(x)$ and explain the result geometrically.

3.6 The Rules of Differentiation

Theorem 1   *Assume that $f(x)$ and $g(x)$ are differentiable at $x = a$. Then:*

(a) *the sum $s(x) = f(x) + g(x)$ is differentiable at $x = a$, and*

$$s'(a) = f'(a) + g'(a); \tag{3.10}$$

(b) *the product $p(x) = f(x) \cdot g(x)$ is differentiable at $x = a$, and*

$$p'(a) = f'(a)g(a) + f(a)g'(a); \tag{3.11}$$

(c) *if $g(a) \neq 0$ the quotient $q(x) = f(x)/g(x)$ is differentiable at $x = a$, and*

$$q'(a) = \frac{f'(a)g(a) - g'(a)f(a)}{(g(a))^2}. \tag{3.12}$$

Proof   We discuss the proof of (b), leaving (a) and (c) as exercises. For (b) we have

$$\lim_{x \to a} \frac{f(x)g(x) - f(a)g(a)}{x - a} = \lim_{x \to a} \frac{f(x)g(x) - f(a)g(x) + f(a)g(x) - f(a)g(a)}{x - a}$$

$$= \lim_{x \to a} g(x) \cdot \lim_{x \to a} \frac{f(x) - f(a)}{x - a} + f(a) \lim_{x \to a} \frac{g(x) - g(a)}{x - a}$$

$$= g(a)f'(a) + f(a)g'(a),$$

by the elementary properties of limits (Theorem 2, Section 3.2), and from the fact that $g(x)$ is continuous at $x = a$ (Theorem 1, Section 3.5). This proves (b).   ∎

Theorem 2 (*The chain rule*)   *Suppose that $f(x)$ is differentiable at $x = a$, and $g(t)$ is differentiable at $t = f(a)$. Then the composite function $h(x) = g(f(x))$ is differentiable at $x = a$, and*

$$h'(a) = g'(f(a)) \cdot f'(a). \tag{3.13}$$

Proof   From the definition of the derivative (Definition 1 of Section 3.5) it follows that $f(x)$ is differentiable at $x = a$ if and only if there exists a number m such that

$$f(x) - f(a) = m(x - a) + u(x) \tag{3.14}$$

where $u(x) = o(x - a)$ as $x \to a$. When this holds, we have $m = f'(a)$.
    Applying this to $g(t)$, we also have

$$g(t) - g(f(a)) = g'(f(a))(t - f(a)) + v(t) \tag{3.15}$$

where $v(t) = o(t - f(a))$ as $t \to f(a)$. Substituting from (3.14) into (3.15), we obtain

$$g(f(x)) - g(f(a)) = g'(f(a))f'(a)(x - a) + w(x)$$

where

$$w(x) = g'(f(a))u(x) + v(f(x)).$$

The desired result, Equation (3.13), will follow if we prove that

$$w(x) = o(x - a) \qquad \text{as } x \to a. \tag{3.16}$$

Let $\varepsilon > 0$ be given. If $g'(f(a)) \neq 0$, choose $\delta_1 > 0$ such that

$$|u(x)| < \frac{\varepsilon}{2|g'(f(a))|} |x - a| \quad \text{if } |x - a| < \delta_1.$$

If $g'(f(a)) = 0$, let $\delta_1 = 1$. Next choose $\delta_2 > 0$ such that

$$|v(f(x))| < \frac{\varepsilon}{2(|f'(a)| + 1)} |f(x) - f(a)| \quad \text{if } |f(x) - f(a)| < \delta_2.$$

By continuity of $f(x)$ at $x = a$, choose $\delta_3 > 0$ such that

$$|f(x) - f(a)| < \delta_2 \quad \text{if } |x - a| < \delta_3.$$

Finally choose δ_4 such that

$$|f(x) - f(a)| < (|f'(a)| + 1)|x - a| \quad \text{if } |x - a| < \delta_4.$$

If

$$\delta = \min(\delta_1, \delta_3, \delta_4),$$

we then have

$$|w(x)| \leq |g'(f(a))| \, |u(x)| + |v(f(x))| < \frac{\varepsilon}{2}|x - a| + \frac{\varepsilon}{2}|x - a| = \varepsilon|x - a|$$

$$\text{if } |x - a| < \delta.$$

This proves (3.16).   ∎

Definition 1   A function $f(x)$ defined on an interval I is said to be **one-to-one** (or **injective**) on I if

$$f(x_1) = f(x_2) \quad \text{implies} \quad x_1 = x_2 \quad \text{for all } x_1, x_2 \in I.$$

Any one-to-one continuous function $f(x)$ must be *strictly monotone* on I (either increasing or decreasing); this intuitively obvious fact will be proved as an exercise in Chapter 4.

Definition 2   Let $f(x)$ be defined and one-to-one on $[a, b]$. Let $g(t)$ be defined on the range† of f and satisfy

$$g(f(x)) = x \quad \text{for all } x \in [a, b]. \tag{3.17}$$

Then $g(t)$ is called the **inverse function** to $f(x)$ on $[a, b]$.

---

† The *range* of $f(x)$ on $[a, b]$ is defined to be the set of all values of $f(x)$, for $x \in [a, b]$.

Writing $t = f(x)$, we see that the inverse function satisfies $x = g(t)$, and therefore

$$f(g(t)) = f(x) = t;$$

in other words, if $g(t)$ is the inverse function to $f(x)$, then $f(x)$ is the inverse function to $g(t)$.

Some well-known examples of inverse functions are

(i) $g(t) = \sqrt{t}$ is the inverse to $f(x) = x^2$ on $[0, \infty)$;

(ii) $g(t) = \log t$ is inverse to $f(x) = e^x$ on $(-\infty, \infty)$;

(iii) $g(t) = \arcsin t$ is inverse to $f(x) = \sin x$ on $\left[-\dfrac{\pi}{2}, \dfrac{\pi}{2}\right]$.

These examples illustrate the fact that the interval of definition of a given elementary function $f(x)$ must often be restricted in order to obtain a one-to-one function, for which the inverse function can then be defined.

The following rule on the derivative of an inverse function is an immediate corollary of Theorem 2.

Corollary *Let $f(x)$ be one-to-one on $[a, b]$ and differentiable on (a, b), and let $g(t)$ be its inverse function. Let $a < x_o < b$. Assume that $f'(x_o) \neq 0$, and that $g(t)$ is differentiable†at $t_o = f(x_o)$. Then we have*

$$g'(t_o) = \frac{1}{f'(x_o)}. \tag{3.18}$$

Exercises

1. Prove Theorem 1(a).

2. Prove the following lemma:

    Let $g(x)$ be differentiable at $x = a$, and assume $g(a) \neq 0$. Then $r(x) = 1/g(x)$ is also differentiable at $x = a$, and

    $$r'(a) = -\frac{g'(a)}{(g(a))^2}.$$

    (You will first have to argue that $g(x) \neq 0$ for x sufficiently close to a.)

3. Show that Theorem 1(c) follows from the above lemma and Theorem 1(b).

4. Use mathematical induction to prove that

    $$\frac{d}{dx}(x^n) = nx^{n-1} \quad \text{for } n = 1, 2, 3, \ldots$$

---

†It can be proved that $g(t)$ is necessarily differentiable at t_o, but the proof requires material from Chapter 4 (Section 4.5, Exercise 8).

5. Use Exercise 4 and the theorems of this section to prove that

$$\frac{d}{dx} x^a = ax^{a-1} \quad (x > 0)$$

for any *rational* number a.

6. If $h(x) = g(f(x))$, show that (under suitable differentiability assumptions)

$$h''(x) = g''(f(x))(f'(x))^2 + g'(f(x))f''(x).$$

4 *Properties of Continuous Functions*

4.1 *Introduction*

In Chapter 1 we stressed the importance of the *completeness property* of the real number system $\mathbb{R}$ in studying the convergence of sequences and series. In this chapter we show how the completeness property is used to prove certain basic theorems about continuous functions. For this purpose, and for many other applications in analysis, it is necessary to consider various other properties of $\mathbb{R}$ which are closely related to completeness. Such properties are discussed in Sections 4.2, 4.3, and 4.4.

In the remaining sections we prove some basic theorems about continuous functions, such as the theorem that *a continuous function defined on a closed bounded interval, $a \leq x \leq b$, has a maximum value* (see Section 4.5). This result is intuitively "obvious" to most students, but present-day standards of mathematical rigor demand strictly logical proofs even for "obvious" results. There are several good reasons for insisting on absolute rigor in mathematics, especially in the foundations. One reason is that seemingly obvious results sometimes turn out to be wrong or, more frequently, imprecise. When more complex situations are studied (for example, calculus in several dimensions), things may no longer be so obvious. Thus we may be forced to analyze the simplest case in great detail before attacking the more complex case. An innocent error at the initial level may become serious when compounded in a complex problem.

The need for absolute, logical precision in mathematics has been recognized since Greek times as a characteristic aspect of the subject. Logical standards have in fact become steadily more rigorous throughout the history of mathematics. As a consequence, mathematics is today the least controversial of all subjects studied by man. Working scientists virtually *never* find it necessary to question the correctness of the mathematics which they use. (Scientists do sometimes complain that mathematicians are too slow in developing the mathematics they require. Unfortunately the required mathematics often happens to be rather difficult.)

The theorem stated earlier—that a continuous function on $a \leq x \leq b$ has a maximum value—is an example of what mathematicians call an *existence theorem*. It tells us that a maximum value *exists*, but not how to find it. The importance of existence theorems can be demonstrated by the following simple example.

Example

Suppose we wish to prove: "1 is the largest positive integer."

Let x denote the largest positive integer. Then $x \geq 1$, so that $x^2 \geq x$. But x^2 is also a positive integer. Therefore $x^2 = x$. Dividing by x, we obtain $x = 1$.

What is the error in this "proof"? The moral of this example is that if we refer to nonexistent objects as if they existed, we may be led into foolish errors. Mathematicians seem to have learned this moral; politicians probably never will.

Exercises

1. Let $x_1 = 2$ and $x_{n+1} = \dfrac{1 + x_n^2}{2}$, for $n \geq 1$. Let $\lim_{n \to \infty} x_n = a$. Then $2a = 1 + a^2$, so that $a = 1$. What is wrong with this?

2. Criticize the following "ontological" proof of the existence of God:

    (a) By definition, God is the greatest conceivable being;

    (b) If God did not exist, then it would be possible to conceive of a greater being (that is, an existent being). This is a contradiction, so God must exist.

3. "Prove" that 1 is the smallest positive real number—and then criticize your "proof."

4.2 Supremum and Infimum

In this section we discuss the important concepts of the supremum (sup) and infimum (inf) of a set A of real numbers. Under special circumstances (to be explained) the supremum of A is the same as the maximum of A, and the infimum of A is the same as the minimum of A.

Definition 1   *Let A be a given subset of* $\mathbb{R}$. *A number x is called an* **upper bound** *of A (respectively, a* **lower bound** *of A) if* $x \geq a$ *for every* $a \in A$ *(respectively,* $x \leq a$ *for every* $a \in A$).

Definition 2   *Let A be a subset of* $\mathbb{R}$. *A number M is called the* **supremum** *(or* **least upper bound**) *of A if*

   (i)   *M is an upper bound of A; and*

   (ii) *if x is any upper bound of A, then we have* $M \leq x$.

Similarly, m is the **infimum** *(or* **greatest lower bound**) *of A if*

   (i)   *m is a lower bound of A; and*

   (ii) *if y is any lower bound of A, then we have* $m \geq y$.

The standard notation is:

$$M = \sup A, \qquad m = \inf A;$$

the notation $M = \text{lub } A$ (for least upper bound of A) and $m = \text{glb } A$ (for greatest lower bound of A) is also common. If A has no upper bound, we write

$$\sup A = +\infty.$$

Similarly if A has no lower bound, we write $\inf A = -\infty$.

Example 1

Let $A = \left\{ \dfrac{x+1}{x} \,\middle|\, x > 0 \right\}$.† Find $\inf A$ and $\sup A$.

Solution   Since $\dfrac{x+1}{x} = 1 + \dfrac{1}{x} \geq 1$, we see that 1 is a lower bound for A. Clearly no number $y > 1$ can be a lower bound for A (why not?). Hence, $\inf A = 1$. Since A contains arbitrarily large numbers, $\sup A = +\infty$.

---

† The notation $\{\, y \mid \cdots \,\}$ stands for *the set of all y such that* . . . .

Example 2

Let $B = \left\{ \dfrac{x+1}{x} \;\middle|\; x \text{ is a positive integer} \right\}$. The reader can easily verify that sup

$B = 2$ and inf $B = 1$.

As an exercise the reader should verify that the supremum of a set A is unique. We will see later that the supremum of A always exists (if the "values" $+\infty$ and $-\infty$ are allowed); this is in fact a consequence of the completeness property of $\mathbb{R}$.

Next let us clarify the relationship between supremum and maximum. Intuitively the *maximum* of a set A of real numbers refers to the largest number in A—*if such a number exists*. Similarly the minimum of A is the smallest number in A—if such a number exists.

Definition 3   *Let A be a subset of $\mathbb{R}$. A real number Q is called the **maximum** of A (notation $Q = \max A$, or $Q = \max_{a \in A} a$) if*

(i) $Q \in A$; *and*

(ii)   *for every $a \in A$ we have $a \le Q$.*

*Similarly, q is called the **minimum** of A (notation $q = \min A$, or $q = \min_{a \in A} a$) if*

(i) $q \in A$

(ii)   *for every $a \in A$ we have $a \ge q$.*

Example 1 above illustrates the fact that a set A may fail to have either a maximum or a minimum; $q = 1$ is *not* the minimum of $A = \left\{ \dfrac{x+1}{x} \;\middle|\; x > 0 \right\}$ (why?).

Theorem 1   *Let A be a subset of $\mathbb{R}$, and let $M = \sup A$. Then A has a maximum, namely M, if and only if $M \in A$.*

Proof   Suppose first that $M \in A$. By Definition 2, M is an upper bound of A, i.e. $M \ge a$ for every $a \in A$. Thus (i) and (ii) of Definition 3 are both satisfied, so that $M = \max A$.

Conversely, suppose $Q = \max A$ exists. Then Q is an upper bound for A (check!). If x is any upper bound of A, we have $x \ge a$ for all $a \in A$, and in particular $x \ge Q$. Hence both (i) and (ii) of Definition 2 are satisfied by Q. Hence (by uniqueness of the supremum), $M = Q$ so that $M \in A$. ∎

To summarize: if a set A has a maximum element Q then Q is also the supremum of A. However, A may have a finite supremum M but no maximum element.

In most applications in this text, the set A will be the set of values of some function: $A = \{f(x) \mid x \in B\}$, where B is some subset of the domain of f. In this case we use the notation

$$\sup_{x \in B} f(x) = \sup \{f(x) \mid x \in B\}.$$

Example 3

If $f(x) = \dfrac{1}{1 + x^2}$, find

$$\sup_{x \in \mathbb{R}} f(x), \quad \inf_{x \in \mathbb{R}} f(x), \quad \max_{x \in \mathbb{R}} f(x), \quad \text{and} \quad \min_{x \in \mathbb{R}} f(x)$$

(if these exist).

Solution It is easy to see that

$$\sup_{x \in \mathbb{R}} \frac{1}{1 + x^2} = \max_{x \in \mathbb{R}} \frac{1}{1 + x^2} = 1$$

and

$$\inf_{x \in \mathbb{R}} \frac{1}{1 + x^2} = 0;$$

but this is not assumed, so that $\min_{x \in \mathbb{R}} \dfrac{1}{1 + x^2}$ does not exist. See Figure 4.1.

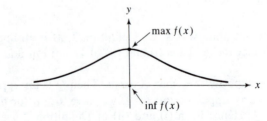

Figure 4.1 $f(x) = 1/(1 + x^2)$.

We remark that in general $M = \max_{x \in B} f(x)$ if and only if (i) $f(x) \leq M$ for all $x \in B$ and (ii) there exists a point $x_0 \in B$ such that $f(x_0) = M$.

Example 3 shows that a function may be bounded and still not have both a maximum and a minimum value. It seems reasonable however that for any bounded function $f(x)$ both $\sup_{x \in B} f(x)$ and $\inf_{x \in B} f(x)$ will exist. This is indeed the case, and the completeness property of $\mathbb{R}$ is required in the proof.

Theorem 2   *Let A be a nonempty subset of* $\mathbb{R}$ *and suppose A has an upper bound. Then* sup *A exists (and is finite).*

This theorem can be reworded as follows: every nonempty subset A of $\mathbb{R}$ which has an upper bound, has a least upper bound. In many texts this statement is assumed as an axiom about the real number system. It can be shown (see Exercise 14) that this statement is logically equivalent to the completeness property of $\mathbb{R}$. Thus it is simply a matter of taste which statement is used as an axiom. In fact, as we mentioned in Chapter 1, both statements can be *proved* as theorems in a more painstaking development of the real number system.

Proof   We will construct a nondecreasing sequence $\{x_n\}$ of points of A, and a nonincreasing sequence $\{y_n\}$ of upper bounds of A, in such a way that

$$\lim_{n \to \infty} x_n = \lim_{n \to \infty} y_n = \lambda. \qquad (4.1)$$

Then we will show that $\lambda = \sup A$.

To begin, let x_1 be any point of A and let y_1 be an upper bound of A; both x_1 and y_1 exist by assumption. Obviously $x_1 \leq y_1$. Let

$$z_1 = \tfrac{1}{2}(x_1 + y_1),$$

so that z_1 is half way between x_1 and y_1. Determine x_2 and y_2 as follows. If z_1 is an upper bound of A, let $x_2 = x_1$ and $y_2 = z_1$. If z_1 is not an upper bound of A, let x_2 be some point of A with $x_2 > z_1$, and let $y_2 = y_1$. In either case we have:

(a) $x_2 \in A$ and y_2 is an upper bound of A;

(b) $x_1 \leq x_2 \leq y_2 \leq y_1$;

(c) $y_2 - x_2 \leq \tfrac{1}{2}(y_1 - x_1)$.

Having determined x_2 and y_2, we can repeat the process to obtain points x_3, y_3. Repeating the process indefinitely (by induction), we obtain two sequences:

a bounded, nondecreasing sequence $\{x_n\}$ of points of A and a bounded, non-increasing sequence $\{y_n\}$ of upper bounds of A. By the completeness property, both of these sequences must converge. Since clearly

$$y_n - x_n \le \frac{1}{2^{n-1}} (y_1 - x_1) \to 0$$

as $n \to \infty$, the sequences must have the same limit λ. This proves Equation (4.1).

We show finally that $\lambda = \sup A$. First let $x \in A$. Then $y_n \ge x$ for all n because each y_n is an upper bound for A. Therefore $\lambda = \lim y_n \ge x$. This shows that λ is an upper bound for A. Next, if $\mu < \lambda$, then $x_n > \mu$ for large n since $\lim x_n = \lambda$. Since $x_n \in A$, this shows that μ is not an upper bound for A. Therefore λ is the least upper bound of A: $\lambda = \sup A$. ∎

Corollary   *Let $f(x)$ be a function defined for all $x \in B$, a nonempty set. Then $\sup_{x \vee B} f(x)$ always makes sense, either as a finite number, or as $+\infty$. Similarly, $\inf_{x \in B} f(x)$ is either finite or $-\infty$.*

Definition 4   *If $\sup_{x \in B} f(x) < \infty$ we say that $f(x)$ is **bounded above** on B, whereas if $\inf_{x \in B} f(x) > -\infty$ we say that $f(x)$ is **bounded below** on B. If $f(x)$ is bounded both above and below on B, we say simply that $f(x)$ is **bounded** on B.*

Exercises

1. Find $\sup A$ and $\inf A$ for each of the following sets A. Determine whether $\max A$ and $\min A$ exist.

    (a) $A = \{x^2 \mid x \in \mathbb{R}\}$.

    (b) A is the finite set $\left\{1, \dfrac{7\pi}{22}, \pi^2, 10\right\}$ (*Note:* $3.1415 < \pi < 3.1416$).

    (c) A is the open interval $(0, 1) = \{x \mid 0 < x < 1\}$.

    (d) A is the half-open interval $(0, 1] = \{x \mid 0 < x \le 1\}$.

    (e) A is the set of all real numbers in $(0, 1)$ whose decimal expansion contains no 9's.

2. (a) Let $\emptyset$ denote the *empty set*, that is, the set that has no elements. What numbers are upper bounds for $\emptyset$? Does $\sup \emptyset$ exist?

(b) By defining sup $\varnothing$ and inf $\varnothing$ appropriately, show that *every* subset of $\mathbb{R}$ has a supremum and an infimum, provided that the "values" $\pm\infty$ are permitted.

3. Find the following sups and infs. State which are maxs and which mins. (Use calculus if necessary; a sketch of the function may also help.)

(a) $\displaystyle \sup_{x>0} \frac{1}{1+x}$,

(b) $\displaystyle \inf_{x>0} \frac{1}{1+x}$,

(c) $\displaystyle \inf_{x>0} \left(x + \frac{1}{x} \right)$,

(d) $\displaystyle \sup_{x \in \mathbb{R}} \frac{x}{\sqrt{1+x^2}}$ and $\displaystyle \inf_{x \in \mathbb{R}} \frac{x}{\sqrt{1+x^2}}$,

(e) $\displaystyle \sup_{x \neq 0} \frac{\sin x}{x}$,

(f) $\displaystyle \inf_{x \neq 0} \left| \frac{\sin x}{x} \right|$,

(g) $\displaystyle \inf_{x \neq 0} \frac{\sin x}{x}$ (You may not be able to solve this exactly!).

4. Explain why sup A is unique if it exists.

5. What if sup $A = $ inf A?

6. Let $A = \{ n^{1/n} \mid n \text{ is a positive integer} \}$. Find sup A and inf A.

7. (a) Let $f(x, y) = x + 2y$. Find

$$\sup_{0 \le x \le 1} \left[\inf_{1 \le y \le 3} f(x, y) \right] \quad \text{and} \quad \inf_{1 \le y \le 3} \left[\sup_{0 \le x \le 1} f(x, y) \right].$$

(b) Show that in general

$$\sup_{x \in A} \left[\inf_{y \in B} f(x, y) \right] \le \inf_{y \in B} \left[\sup_{x \in A} f(x, y) \right].$$

(*Hint:* Begin with the statement that $f(x, y) \le \sup_{x \in A} f(x, y)$ for all $x \in A$, $y \in B$.)

(c) Show by example that the inequality in part (b) may be strict. (Consider the following picture.)

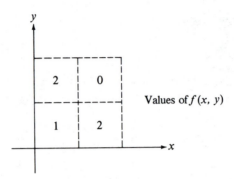

Values of $f(x, y)$

8. Prove that $M = \sup A$ if and only if
    (a) M is an upper bound of A, and
    (b) there exists a sequence $\{x_n\}$ in A with $\lim_{n \to \infty} x_n = M$.

9. Prove that for $A \neq \varnothing$

    $$\sup_{x \in A} [f(x) + g(x)] \leq \sup_{x \in A} f(x) + \sup_{x \in A} g(x). \tag{4.2}$$

    (*Hint:* Let $M = \sup_{x \in A} f(x)$ and $N = \sup_{x \in A} g(x)$. Then $f(x) \leq M$ and $g(x) \leq N$ for every $x \in A$. Complete this argument.)

10. Give examples of functions $f(x)$ and $g(x)$ such that

    $$\sup_{x \in [0,1]} [f(x) + g(x)] < \sup_{x \in [0,1]} f(x) + \sup_{x \in [0,1]} g(x);$$

    also give examples in which this is an equality.

11. State and prove the analogous inequality to (4.2) for the case of the infimum.

12. Let $f(x)$ be a nondecreasing function defined for $a \leq x \leq b$. Show that the right- and left-hand limits of $f(x)$ exist at every point z with $a < z < b$.

*13. Let $f(x)$ be a nondecreasing function defined for $a \leq x \leq b$. Show that the points at which $f(x)$ is discontinuous are either finite or countable. (*Hint:*

For each positive integer n let S_n denote the set of all points z in $[a, b]$ such that

$$\lim_{x \to z+} f(x) - \lim_{x \to z-} f(x) > \frac{1}{n};$$

each S_n must be a finite set.)

*14. Assume that $\mathbb{R}$ satisfies conditions 1–10 of Section 1.3, as well as

    11'. *Every nonempty subset A of $\mathbb{R}$ that has an upper bound has a least upper bound.*

    Prove that $\mathbb{R}$ has the completeness property. (*Hint:* If $\{x_n\}$ is a nondecreasing bounded sequence, then $\lambda = \sup_{n \geq 1} x_n$ exists. Show that $\lambda = \lim_{n \to \infty} x_n$.)

4.3 The Bolzano-Weierstrass Property

The completeness property tells us that a bounded *monotonic* sequence must converge. In this section we show that *every* bounded sequence must have some convergent *subsequence*.

Definition   *Let S be a given subset of $\mathbb{R}$. A point $x_0 \in \mathbb{R}$ is called an accumulation point of S provided that every deleted neighborhood of x_0 contains some point of S.*

It can easily be shown that x_0 is an accumulation point of S if and only if there is some sequence $\{x_n\}$ in S such that $\lim_{n \to \infty} x_n = x_0$ and $x_n \neq x_0$ for every n (see Exercise 4). You should verify the following simple examples:

(a) A finite set has no accumulation points.
(b) If $S = (0, 1)$, then the set of all accumulation points of S is the closed interval $[0, 1]$.
(c) If $S = \mathbb{Q}$ (the rationals), then every real number is an accumulation point of S.

Theorem (Bolzano-Weierstrass)   *A bounded, infinite set of real numbers has at least one accumulation point.*

Proof Let S be a given bounded, infinite set; then there is some interval $[a_1, b_1]$ containing S. Let

$$c_1 = \tfrac{1}{2}(a_1 + b_1).$$

Since S is an infinite set, at least one of the intervals $[a_1, c_1]$ or $[c_1, b_1]$ contains infinitely many points of S. Let $[a_2, b_2]$ denote such an interval. Note that

$$a_1 \leq a_2 \leq b_2 \leq b_1 \quad \text{and} \quad b_2 - a_2 = \tfrac{1}{2}(b_1 - a_1).$$

The same process may be repeated inductively to obtain a sequence of intervals $[a_n, b_n]$, each containing infinitely many points of S, and such that

$$a_1 \leq a_2 \leq a_3 \leq \cdots \leq a_n \leq \cdots \leq b_n \leq \cdots \leq b_2 \leq b_1 \qquad (4.3)$$

and

$$b_n - a_n = \frac{1}{2^{n-1}}(b_1 - a_1). \qquad (4.4)$$

According to the completeness property, $a = \lim_{n \to \infty} a_n$ and $b = \lim_{n \to \infty} b_n$ exist; (4.4) obviously implies that $a = b$. We assert that a is an accumulation point of S. To see this, suppose $\varepsilon > 0$ is given. Then for sufficiently large n we have

$$a - \varepsilon < a_n \leq b_n < a + \varepsilon.$$

Since $[a_n, b_n]$ contains infinitely many points of S according to the construction, the interval $(a - \varepsilon, a + \varepsilon)$ must surely contain some point of S not equal to a. This completes the proof. ∎

Notice that the proof of the Bolzano-Weierstrass theorem once again used the completeness property of $\mathbb{R}$. Conversely, it is possible to prove the completeness property of $\mathbb{R}$ if one assumes the validity of the Bolzano-Weierstrass theorem; see Exercise 5.

Corollary *Let $\{x_n\}$ be a sequence of real numbers. If $\{x_n\}$ is bounded, then $\{x_n\}$ has a convergent subsequence.*

Proof Let S denote the set of all values x_n, $n = 1, 2, 3, \ldots$. There are two cases to consider: either S is a finite set or S is an infinite set.

In case S is a finite set, there must be infinitely many values of n, say $n_1, n_2, n_3, \ldots$, such that $x_{n_k} = x$ where x is one of the members of S. Hence, $\{x_n\}$ has a convergent subsequence.

If S is an infinite set, we apply the Bolzano-Weierstrass theorem to obtain an accumulation point a of S. It follows (cf. Exercise 4) that some subsequence of $\{x_n\}$ must converge to a. ∎

Exercises

1. Find the set of all accumulation points for each of the following sets

   (a) $S = \{x \in \mathbb{R} \mid 0 < |x| < 1\}$.
   (b) S is the set of all integers.
   (c) S is the set of all numbers $m/2^n$, where m, n are integers.
   (d) S is the set of all rational numbers p/q in which $|p| < 10$.

2. Let S be a bounded infinite set in $\mathbb{R}$. Show that either $\sup S = \max S$ or $\sup S$ is an accumulation point of S. Give examples where one of these possibilities holds but not the other. Give an example where both hold.

3. If $\{x_n\}$ is a given sequence, let S denote the set of values of x_n, $n = 1, 2, 3, \ldots$, and let S' denote the set of accumulation points of S. Find examples of sequences $\{x_n\}$ for which

   (a) S' is empty,
   (b) S' consists of one point,
   (c) S' consists of two points,
   (d) S' is countable,
   (e) S' is uncountable.

4. Prove that x is an accumulation point of a set S if and only if there is a sequence $\{x_n\}$ of points of S such that $x_n \neq x$ for all n, and $\lim_{n \to \infty} x_n = x$.

5. Assuming that the real number system $\mathbb{R}$ has Properties 1–10 of Section 1.3, and that the Bolzano-Weierstrass theorem is valid, prove that the completeness property holds in $\mathbb{R}$.

4.4 Cauchy Sequences

We often wish to know whether a given sequence $\{x_n\}$ converges. Sometimes we can calculate $\lim_{n \to \infty} x_n$ by inspection or by using simple theory. If $\{x_n\}$ is increasing or decreasing we can appeal to the completeness property. But many other sequences can arise. For many years mathematicians tried to find a necessary and sufficient condition for the convergence of a given sequence $\{x_n\}$. Some people asserted that $\lim_{n \to \infty} (x_{n+1} - x_n) = 0$ was such a condition, but we know this is false (cf. Exercise 1). The problem was solved by A. Cauchy.

Definition   *A sequence $\{x_n\}$ of real numbers is called a **Cauchy sequence** if it satisfies the following condition:*

> *for any given $\varepsilon > 0$ there exists an integer N (which may depend on ε) such that*

$$|x_n - x_m| < \varepsilon \quad \text{for all } n, m \geq N. \tag{4.5}$$

Notice that although this definition is reminiscent of the definition of the limit of a sequence, it refers *only* to the terms of the sequence $\{x_n\}$, and not to any anticipated limit L.

Theorem (*Cauchy*)   *A necessary and sufficient condition for convergence of a sequence $\{x_n\}$ is that it be a Cauchy sequence.*

Proof   The proof of necessity is very easy and will be given first. Suppose that $\lim_{n \to \infty} x_n = L$ exists. Let $\varepsilon > 0$ be given, and choose an integer N such that $|x_n - L| < \varepsilon/2$ if $n \geq N$. Then, for both $n, m \geq N$, we have

$$|x_n - x_m| \leq |x_n - L| + |x_m - L| < \varepsilon,$$

which shows that $\{x_n\}$ is a Cauchy sequence.

The converse proof is slightly more difficult. First we will show that a Cauchy sequence is bounded. For this purpose let $\varepsilon = 1$; then there exists an integer N such that $|x_n - x_m| < 1$ for $n, m \geq N$. In particular $|x_n - x_N| < 1$ for $n \geq N$, which shows that the sequence $\{x_n\}$ is bounded for $n \geq N$, and hence for all n.

Now by the corollary to the Bolzano-Weierstrass theorem, the bounded sequence $\{x_n\}$ must have a convergent subsequence $\{x_{n_k}\}$. Let $\lim_{k \to \infty} x_{n_k} = L$. For any integer n we have

$$|x_n - L| \leq |x_n - x_{n_k}| + |x_{n_k} - L|.$$

Given $\varepsilon > 0$ we can find an integer N such that $|x_n - x_m| < \varepsilon/2$ for all $n, m \geq N$, and also $|x_{n_k} - L| < \varepsilon/2$ for $k \geq N$. Therefore, if $n \geq N$, we can choose a term x_{n_k} such that

$$|x_n - L| \leq |x_n - x_{n_k}| + |x_{n_k} - L| < \frac{\varepsilon}{2} + \frac{\varepsilon}{2} = \varepsilon.$$

This shows that $\{x_n\}$ converges to L.   ∎

The condition (4.5) is often called the *Cauchy criterion* for convergence; it is sometimes written as

$$\lim_{n,m \to \infty} (x_n - x_m) = 0. \tag{4.6}$$

For infinite series Cauchy's theorem immediately implies a corollary.

Corollary   *An infinite series $\sum_1^\infty a_n$ converges if and only if*

$$\lim_{n,m \to \infty} \sum_{k=n}^m a_k = 0,$$

that is, if and only if :

for every $\varepsilon > 0$ there exists an integer N (which may depend on ε) such that

$$|a_n + a_{n+1} + \cdots + a_m| < \varepsilon \quad \text{whenever } m \geq n \geq N.$$

Exercises

1. Show that if $\{x_n\}$ converges, then $\lim_{n \to \infty} (x_{n+1} - x_n) = 0$, *but not conversely.*

2. Suppose that $|x_{n+1} - x_n| \leq \dfrac{1}{2^n}$ for all n. Prove that $\{x_n\}$ converges. Can you generalize this result?

3. Let $f(x)$ be defined for $x > 0$. Show that $\lim_{x \to \infty} f(x)$ exists (finitely) if and only if for every given $\varepsilon > 0$ there exists a number M such that

    $$|f(x) - f(y)| < \varepsilon \quad \text{whenever } x \geq M \text{ and } y \geq M.$$

    (*Hints:* Necessity is easy. For sufficiency, apply Cauchy's theorem to the sequence $\{f(n)\}$ to prove that $\lim_{n \to \infty} f(n) = L$ exists. You still have to show that $\lim_{x \to \infty} f(x) = L$.)

4. An infinite series $\sum_1^\infty a_n$ is said to converge *absolutely* if the series $\sum_1^\infty |a_n|$ converges. Use the corollary in this section to give a simple proof that absolute convergence implies ordinary convergence.

5. Prove that if $\{x_n\}$ is a sequence decreasing to zero then $\sum_1^\infty |x_{n+1} - x_n|$ converges (this is trivial!), but that the same series may diverge if $x_n \to 0$ but not monotonically.

4.5   *Properties of Continuous Functions*

In this section we prove several basic theorems about continuous functions. Although these theorems are often referred to in calculus texts, their proofs are usually considered "beyond the scope" of such texts.

In each of the theorems, it will be supposed that f is a function that is defined and continuous on a closed, finite interval $[a, b]$.† Remember that by definition this means that f is continuous at each point x_0 in $[a, b]$. Strictly speaking we have not yet defined continuity at an end point. This omission is easily rectified: when f is defined on a closed interval $[a, b]$ we say that f is continuous at a provided

$$\lim_{x \to a+} f(x) = f(a),$$

and similarly f is *continuous at b* provided

$$\lim_{x \to b-} f(x) = f(b).$$

Example

The function $f(x) = \sqrt{1 - x^2}$ is continuous on the closed interval $[-1, 1]$.

Theorem 1 (*Intermediate value theorem*)   *Let f be defined and continuous on the closed finite interval* $[a, b]$. *Suppose that*

$$f(a) < 0 < f(b).$$

Then there exists a point c in the open interval (a, b) *such that* $f(c) = 0$ (*see Figure 4.2*).

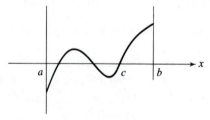

Figure 4.2   The intermediate value theorem.

---

† Henceforth we adopt the modern notation for functions. A single symbol (f, g, Ψ, etc.) will denote a *function;* the notation $f(x)$ will be used to denote the *value* of the function f at the point x. (Nevertheless we will also continue to use expressions such as "the function $f(x) = x^2$" rather than the pedantic form "the function f whose value at x is x^2.")

Proof The idea of the proof is to define c as the "largest" value of x for which $f(x) \leq 0$, namely,

$$c = \sup \{x \in [a, b] \,|\, f(x) \leq 0\}.$$

That such a number c exists follows from Section 4.2: the set $\{x \in [a, b] \,|\, f(x) \leq 0\}$ is nonempty (because $f(a) < 0$) and bounded above by b. We wish to show that $f(c) = 0$. Suppose that $f(c) < 0$. By continuity of f at c there exists $\delta > 0$ such that $f(c + \delta) < 0$. But this contradicts the definition of c.

Similarly if $f(c) > 0$, then there exists $\delta > 0$ such that $f(x) > 0$ for all $x > c - \delta$ (note that $f(x) > 0$ for all $x > c$ by the choice of c). This again contradicts the definition of c. Therefore $f(c) = 0$. Finally, since $f(a) < 0$ and $f(b) > 0$, we cannot have $c = a$ or b, so that $c \in (a, b)$. ∎

In general, if f is continuous on $[a, b]$, then $f(x)$ must assume *every* value between $f(a)$ and $f(b)$ at some point x between a and b. This is easily proved from Theorem 1 (see Exercise 2).

Corollary *Any polynomial of odd degree has at least one real root.*

Proof Let $p(x) = a_0 x^k + \cdots + a_k$ be such a polynomial, where k is odd and, without loss of generality, $a_0 > 0$. We know that $p(x)$ is continuous for all x and also that

$$\lim_{x \to +\infty} p(x) = +\infty \quad \text{and} \quad \lim_{x \to -\infty} p(x) = -\infty.$$

Hence, we can choose some interval $[a, b]$ such that $p(a) < 0 < p(b)$, and therefore $p(x) = 0$ for some point x. ∎

Another corollary is: *every positive real number a has a positive nth root* (see Exercise 6).

We now establish another fundamental property of continuous functions. First we state a useful lemma.

Lemma *Suppose that f is continuous at x_0. Then if $\{x_n\}$ is any sequence converging to x_0, it follows that the sequence $\{f(x_n)\}$ converges to $f(x_0)$.*

The proof is a simple ε-δ argument, which we leave to you to carry out. The converse of this lemma is also valid: if $f(x_n) \to f(x_0)$ for *every* sequence $\{x_n\}$ which converges to x_0, then f is continuous at x_0; see Exercise 9.

Theorem 2   *Let f be defined and continuous on the closed finite interval [a, b].
Then f is bounded on [a, b]; that is,*

$$\sup_{x \in [a,b]} |f(x)| < +\infty.$$

Proof   We prove this by contradiction. Suppose f is not bounded on $[a, b]$;
for example, suppose f is not bounded from above. Then for any positive
integer n there must exist a point x_n in $[a, b]$ such that $f(x_n) \geq n$. According to
the Bolzano-Weierstrass theorem, the sequence $\{x_n\}$ must have a convergent
subsequence $\{x_{n_k}\}$; let $x_0 = \lim_{k \to \infty} x_{n_k}$. Then $x_0 \in [a, b]$. Since f is continuous
at x_0, we must have, by the lemma,

$$\lim_{k \to \infty} f(x_{n_k}) = f(x_0).$$

But since $f(x_{n_k}) \geq n_k$, we have instead

$$\lim_{k \to \infty} f(x_{n_k}) = +\infty.$$

This contradiction completes the proof.   ∎

Theorem 3   *Let f be defined and continuous on the closed finite interval [a, b].
Then $\max_{x \in [a, b]} f(x)$ and $\min_{x \in [a, b]} f(x)$ both exist. (In other words, $f(x)$ assumes
its maximum and minimum values on [a, b].)*

Proof   Let $M = \sup_{x \in [a, b]} f(x)$; by Theorem 2 we know that M is a finite
number. It follows from the definition of the supremum that there is a sequence
$\{x_n\}$ in $[a, b]$ such that $f(x_n) \to M$ as $n \to \infty$ (see Exercise 8, Section 4.2). By
the Bolzano-Weierstrass theorem, this sequence has a convergent subsequence
$\{x_{n_k}\}$. Thus we may write $x_0 = \lim_{k \to \infty} x_{n_k}$. Then

$$f(x_0) = \lim_{k \to \infty} f(x_{n_k}) = M.$$

Therefore $M = \max_{x \in [a, b]} f(x)$. Similarly

$$m = \inf_{x \in [a, b]} f(x) = \min_{x \in [a, b]} f(x).   ∎$$

Theorem 4 *Let f be defined and continuous on the closed interval* [a, b], *and suppose* $f(x) \neq 0$ *for all* $x \in [a, b]$. *Then, either* $f(x) > 0$ *for all* $x \in [a, b]$ *or* $f(x) < 0$ *for all* $x \in [a, b]$.

In case $f(x) > 0$ *for all x in* [a, b], *there exists a positive number* δ *such that* $f(x) \geq \delta > 0$ *for all x in* [a, b].

This theorem is an immediate consequence of Theorems 1 and 3. We leave the details to the reader. Sometimes the theorem is stated briefly as follows: "a continuous function which does not vanish on a closed, finite interval, is bounded away from zero on the interval."

Exercises

1. Let $f(x) = x - [x]$, where $[x]$ denotes the greatest integer in x. Show that although f is bounded, $\max_{[0, 2]} f(x)$ does not exist.

2. Prove the *generalized intermediate value theorem:* if f is continuous on [a, b], then $f(x)$ assumes every value between $f(a)$ and $f(b)$ at some point of [a, b]. (*Hint:* If ξ lies between $f(a)$ and $f(b)$, apply Theorem 1 to the function $f(x) - \xi$.)

3. Show that if f is a continuous function defined on [0, 1] and if $0 \leq f(x) \leq 1$ for all x, the equation $f(x) = x$ has a solution in [0, 1].

4. Let $p(x) = x^4 + a_1 x^3 + a_2 x^2 + a_3 x + a_4$. Show that if $a_4 < 0$, then $p(x)$ has at least two distinct zeros.

5. Show by examples that a continuous function defined on an open, finite interval need not be bounded, and even if bounded, need not have a maximum value on the given interval.

6. Prove that every positive real number a has a unique positive nth root $(n = 2, 3, \ldots)$.

7. Let f be continuous on [a, b], $a < b$. Suppose f is also *one-to-one*, that is, $x_1 \neq x_2$ implies $f(x_1) \neq f(x_2)$. Prove that f must be a monotone function.

*8. Let f be a strictly increasing, continuous function on [a, b]. Then there is a uniquely defined inverse function g defined on the interval $[f(a), f(b)]$ (that is, $g(f(x)) = x$ for all x in [a, b]). Prove that g is continuous.

9. (a) Prove the lemma of this section in detail.

(b) In terms of ε-δ, what does the statement $\lim_{x \to a} f(x) \neq L$ mean? (cf. Appendix I, Section 2.)

(c) Prove the converse described following the lemma.

4.6   Uniform Continuity

We now consider the concept of continuity a little more deeply. Let f be a function defined on an interval I; we allow the possibility of infinite intervals I. We have defined f to be continuous on I if f is continuous at each point of I. From the definition of continuity at a point x, we therefore have: *f is continuous on I if, for every $x \in I$ and for every $\varepsilon > 0$, there exists a number $\delta > 0$ such that*

$$|f(x') - f(x)| < \varepsilon \quad \text{if } |x' - x| < \delta. \tag{4.7}$$

Example 1

Let $f(x) = x^2$ on $I = (-\infty, \infty)$. Calculate δ so that (4.7) is valid.

Solution   First let $\delta \leq 1$. Then $|x' - x| < \delta$ will imply that

$$|x' + x| \leq |x' - x| + |2x| \leq 1 + 2|x|.$$

Therefore

$$|x'^2 - x^2| = |x' - x| \cdot |x' + x|$$

$$\leq |x' - x| \cdot (1 + 2|x|) \qquad \text{(if } |x' - x| \leq 1)$$

$$< \varepsilon \qquad \left(\text{if } |x' - x| < \frac{\varepsilon}{1 + 2|x|} \right).$$

Hence, we may choose

$$\delta = \min \left(1, \frac{\varepsilon}{1 + 2|x|} \right). \tag{4.8}$$

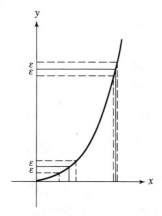

Figure 4.3 $f(x) = x^2$.

Notice that *this value of δ depends on x;* as $|x|$ increases, the value of δ given by (4.8) decreases. Consideration of the graph $y = x^2$ (see Figure 4.3) makes it clear that (for a given ε) the value of δ required in (4.7) *necessarily* decreases as $|x|$ increases. Moreover no fixed value of $\delta > 0$ will work for all x in $(-\infty, \infty)$.

Example 2

If $f(x) = x^2$ is considered only on the *bounded* interval $0 \leq x \leq 3$, then a value δ independent of x can be found so that (4.7) holds.

To see this, we simply note that if $0 \leq x \leq 3$, then $1 + 2\,|x| \leq 7$. Hence, by the same calculations as in Example 1,

$$|f(x') - f(x)| = |x'^2 - x^2| \leq |x' - x|\,(1 + 2\,|x|) \qquad (\text{if } |x' - x| \leq 1)$$

$$\leq 7\,|x' - x|$$

$$< \varepsilon \qquad \left(\text{if also } |x' - x| < \frac{\varepsilon}{7}\right).$$

Hence we may choose $\delta = \min\,(1,\,\varepsilon/7)$, and this is independent of x.

A function f defined on an interval I will be called *uniformly continuous* on I if the value δ in (4.7) exists and can be chosen independently of $x \in I$.

More precisely, we have:

Definition   *Let f be defined on an interval I. Then f is said to be **uniformly continuous** on I if the following condition holds:*

for any given $\varepsilon > 0$, there is a corresponding $\delta > 0$ such that
$$|f(x_1) - f(x_2)| < \varepsilon \quad \text{for all points } x_1, x_2 \in I \text{ with } |x_1 - x_2| < \delta.$$

(4.9)

We emphasize that the number δ may depend on ε, *but not on x_1 or x_2.* (For simplicity of use, this definition has been made symmetric in the values of x appearing in (4.7), which we now denote by x_1, x_2 instead of x, x'.)

We can now prove the following important result.

Theorem   *Let f be defined and continuous on a closed, finite interval I. Then f is uniformly continuous on I.*

Proof   The proof will be by contradiction; suppose f is not uniformly continuous on I. Negating condition (4.9) (cf. Appendix I), we therefore have:

there exists a number $\varepsilon > 0$ such that for every $\delta > 0$ there are points $x, y \in I$ (depending on δ!) such that

$$|x - y| < \delta \quad \text{and} \quad |f(x) - f(y)| \geq \varepsilon.$$

We will apply this statement to the values $\delta = 1, 1/2, 1/3, \ldots, 1/n, \ldots$. For $\delta = 1$ we obtain points $x_1, y_1 \in I$ such that

$$|x_1 - y_1| < 1 \quad \text{and} \quad |f(x_1) - f(y_1)| \geq \varepsilon.$$

In general, for every integer $n \geq 1$, we obtain points $x_n, y_n \in I$ such that

$$|x_n - y_n| < \frac{1}{n} \quad \text{and} \quad |f(x_n) - f(y_n)| \geq \varepsilon.$$

(4.10)

Since I is a closed, finite interval, the Bolzano-Weierstrass theorem implies that the sequence $\{x_n\}$ has a convergent subsequence $\{x_{n_k}\}$. Let $x_0 = \lim_{k \to \infty} x_{n_k}$. Since $|x_{n_k} - y_{n_k}| < 1/n_k \to 0$ as $k \to \infty$, it follows that $\lim_{k \to \infty} y_{n_k} = x_0$ also.

By hypothesis $f(x)$ is continuous at x_0. Therefore

$$\lim_{k \to \infty} f(x_{n_k}) = f(x_0) = \lim_{k \to \infty} f(y_{n_k}).$$

Consequently we can find an integer n_k such that

$$|f(x_{n_k}) - f(y_{n_k})| \le |f(x_{n_k}) - f(x_0)| + |f(y_{n_k}) - f(x_0)|$$

$$< \frac{\varepsilon}{2} + \frac{\varepsilon}{2} = \varepsilon.$$

This contradicts (3.10), and the proof is complete. ∎

Exercises

1. The following functions are uniformly continuous on the indicated intervals. Calculate a value of δ independent of x_1, x_2 such that (4.9) is valid.

    (a) $f(x) = \dfrac{x}{x+1}$; $I = [0, 3]$,

    (b) $f(x) = 2x^2 - x$; $I = [0, 10]$,

    (c) $f(x) = 3x^{-1}$; $I = [1, \infty)$.

2. Show that a linear function $f(x) = ax + b$ is uniformly continuous on any interval I.

3. The function $f(x) = 1/x$ is continuous on the open interval $(0, 1)$. Show that it is not uniformly continuous there.

4. Is $f(x) = |x|$ uniformly continuous on $(-\infty, \infty)$?

5. Let $f(x) = \sin \pi/x$ be defined on $I = (0, 1)$. It should be obvious that f is not uniformly continuous on I. Where does the proof of the Theorem of this section break down?

6. Prove that if f is uniformly continuous on the open, bounded interval $I = (a, b)$, then f is bounded on I. Does it follow that $\max_{x \in I} f(x)$ exists?

7. Let $f(x) = 1/(1 + x^2)$. By determining a number δ independent of x, show that f is uniformly continuous on $(-\infty, \infty)$.

8. Let $f(x)$ be uniformly continuous on (a, b) and let $\{x_n\}$ be a sequence of points in (a, b) converging to a. Prove that $\lim_{n \to \infty} f(x_n)$ exists. (*Hint:* Use Cauchy's theorem, Section 4.4.)

5 *Some Theorems of Calculus*

On the basis of the theorems of Chapter 4 we can now give rigorous proofs of certain basic theorems of elementary calculus.

5.1 *The Mean Value Theorem*

Definition 1   Let f be defined on $[a, b]$. Then f is said to have a **relative maximum** (or **local maximum**) at $c \in [a, b]$ if there exists $\varepsilon > 0$ such that

$$f(x) \le f(c) \quad \text{for all } x \in [a, b] \text{ such that } |x - c| < \varepsilon.$$

The definition of *relative (local) minimum* is similar.

Theorem 1   Let f, defined on $[a, b]$, have a relative maximum or minimum at $c \in (a, b)$. If f is differentiable at $x = c$ then $f'(c) = 0$.

Proof   Suppose, to be explicit, that f has a relative maximum at c. Choose $\varepsilon > 0$ such that $f(x)$ is defined for $|x - c| < \varepsilon$, and $f(x) \le f(c)$ there.

Then we have $f(c + h) - f(c) \le 0$ for all h such that $|h| < \varepsilon$. Consequently

$$\frac{f(c + h) - f(c)}{h} \begin{cases} \le 0 & \text{if } h > 0, \\ \\ \ge 0 & \text{if } h < 0. \end{cases}$$

Since

$$f'(c) = \lim_{h \to 0} \frac{f(c + h) - f(c)}{h}$$

exists by hypothesis, we conclude that both $f'(c) \geq 0$ and $f'(c) \leq 0$. Therefore $f'(c) = 0$. ∎

Note that in the above theorem the point c is not allowed to be an end point of the interval $[a, b]$. However, if for example f has a relative maximum at $x = b$, the proof of Theorem 1 shows that

$$f'(b) \geq 0$$

provided $f'(b)$ exists. See Exercise 2.

We proceed next to the proof of the *mean value theorem* (for derivatives), beginning with the following special case.

Lemma (Rolle's theorem)   *Let f be defined on $[a, b]$, $a \neq b$. Assume that*

 (i) *f is continuous on $[a, b]$,*
 (ii) *f is differentiable on (a, b),*
 (iii) *$f(a) = f(b) = 0$.*

Then there is a point $c \in (a, b)$ for which $f'(c) = 0$.

Proof   (See Figure 5.1.) Since f is continuous on the closed interval $[a, b]$, it has both a maximum value $f(c_1)$ and a minimum value $f(c_2)$ on $[a, b]$. If $f(c_1) = f(c_2) = 0$ we have $f(x) \equiv 0$ on $[a, b]$, so that $f'(c) = 0$ for any $c \in (a, b)$. Otherwise either $f(c_1) > 0$ or $f(c_2) < 0$. For example, if $f(c_1) > 0$ then $c_1 \in (a, b)$; also $f'(c_1) = 0$ by Theorem 1. ∎

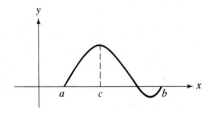

Figure 5.1   Rolle's theorem.

To derive the mean value theorem from Rolle's theorem, we consider a function f satisfying hypotheses (i) and (ii) of Rolle's theorem, but having arbitrary values at $x = a$ and $x = b$. Let $l(x)$ be the *linear* function passing through the points $(a, f(a))$ and $(b, f(b))$; see Figure 5.2. Then

$$l(x) = f(a) + \frac{f(b) - f(a)}{b - a}(x - a). \tag{5.1}$$

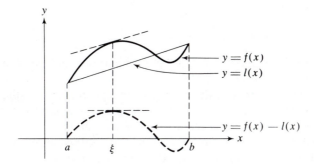

Figure 5.2   The mean value theorem.

Now let $g(x) = f(x) - l(x)$; it is easily seen that $g(x)$ satisfies all three hypotheses of Rolle's theorem. Consequently there is some point $\xi \in (a, b)$ such that

$$g'(\xi) = f'(\xi) - l'(\xi) = f'(\xi) - \frac{f(b) - f(a)}{b - a} = 0.$$

We have therefore proved the following theorem.

Theorem 2 (*The mean value theorem*)   *Let f be defined on* $[a, b]$*, where* $a < b$. *Assume that*

(i) *f is continuous on* $[a, b]$,
(ii) *f is differentiable on* (a, b).

Then there exists some point $\xi \in (a, b)$ *such that*

$$f'(\xi) = \frac{f(b) - f(a)}{b - a}. \tag{5.2}$$

The mean value theorem has the following two corollaries, which calculus students usually consider to be "obvious."†

Corollary 1   *Suppose that f is defined on (a, b) and that $f'(x) \equiv 0$ (that is, $f'(x) = 0$ for every x in (a, b)). Then $f(x)$ is constant on (a, b).*

Corollary 2   *Suppose that f is defined on (a, b) and that $f'(x) > 0$ for all x in (a, b). Then f is strictly increasing on (a, b).*

The proofs *are* fairly "obvious" consequences of the mean value theorem, and are left for the exercises.

We will use the mean value theorem and its corollaries in studying integration in Section 5.3. Before leaving this theorem, let us comment briefly on its hypotheses. Notice that f is required to be continuous on the *closed* interval $[a, b]$ and differentiable on the *open* interval (a, b). If either of these hypotheses is violated, the conclusion fails (see Exercises 8 and 9).

Exercises

1. Find all relative maxima and minima of the following functions (*note:* end points must be considered). Sketch graphs.

   (a) $f(x) = x^3 - x^2$ on $[0, 1]$,
   (b) $f(x) = x^3 - x^2$ on $[-1, 2]$,
   (c) $f(x) = xe^{-x}$ on $[0, 2]$,
   (d) $f(x) = |x^2 - x|$ on $[0, 2]$.

2. Prove that if f has a relative *minimum* on $[a, b]$ at $x = b$, then $f'(b) \leq 0$ if it exists. Illustrate. Also consider the case of the left-hand end point $x = a$. (*Note:* If f is defined only on $[a, b]$ then $f'(b)$ is of course undefined. You should either assume f to be defined on a larger interval than $[a, b]$, or else consider the "one-sided derivative" $f'_-(b) = \lim_{h \to 0-} (f(b + h) - f(b))/h$.)

---

† Professional mathematicians use the word "obvious" to indicate that it is obvious how to give a complete proof. To use "obvious" to mean "I'm sure it's true, but I can't prove it," is not a commendable practice.

3. Let $f(x) = \sqrt{x}$, $0 \le x \le a$. Find all points $\xi \in (0, a)$ for which the mean value formula,

$$f'(\xi) = \frac{f(a) - f(0)}{a - 0} = \frac{f(a)}{a},$$

is valid. Does f satisfy all hypotheses of the mean value theorem?

4. Prove Corollary 1. (*Hint:* If $a < x_0 < x_1 < b$, prove that $f(x_0) = f(x_1)$. This proves the corollary; why?)

5. Prove Corollary 2.

6. Prove: If $f'(x) \ge 0$ on (a, b), then $f(x)$ is nondecreasing on (a, b).

7. State analogues to Corollary 2 and to Exercise 6, for the cases $f'(x) < 0$, $f'(x) \le 0$, respectively. Give one-line proofs of these analogues, without using the word "similarly."

8. Let $f(x) = 1 - |x|$ be defined on $[-1, 1]$. Show that f satisfies neither the hypotheses nor the conclusion of Rolle's theorem.

9. Give an example of a function f which is differentiable on $(0, 1)$, but not continuous on $[0, 1]$, for which the conclusion of the mean value theorem fails.

10. Give an example of a function f satisfying the hypotheses of Rolle's theorem on $[0, 1]$ and having infinitely many points ξ satisfying the conclusion of Rolle's theorem.

11. Use Rolle's theorem to show that the polynomial equation $x^3 + x - 9 = 0$ has exactly one real solution.

12. Show that if $a > 0$ and b is an arbitrary real number, then the equation $x^3 + ax + b = 0$ has exactly one real solution.

13. Show that $|\sin x - \sin y| \le |x - y|$.

14. Prove that if $f'(x) \equiv g'(x)$ on (a, b), then $f(x) \equiv g(x) + \text{const.}$ on (a, b). (Give a very short proof!)

15. Show that $f''(x) \equiv 0$ implies $f(x) \equiv ax + b$.

16. Let f be defined on $(-\infty, \infty)$ and suppose

$$\sup_{-\infty < x < \infty} |f'(x)| < \infty.$$

Show that f is uniformly continuous on $(-\infty, \infty)$.

17. Prove the "generalized mean value theorem": Let f, g be defined on $[a, b]$, where $a < b$. Assume that f, g are continuous on $[a, b]$ and differentiable on (a, b), and that $g'(x) \neq 0$ for $x \in (a, b)$. Then there exists some point $\xi \in (a, b)$ such that

$$\frac{f'(\xi)}{g'(\xi)} = \frac{f(b) - f(a)}{g(b) - g(a)}.$$

(*Hint:* Consider the function $k(x) = [f(b) - f(a)]g(x) - [g(b) - g(a)]f(x)$.)

5.2   The Riemann Integral

The definition of integral given in this section is due to G. Darboux (1842–1917); it is equivalent to a definition given earlier by B. Riemann (1826–1866). The motivation for the definition is, of course, that the definite integral

$$\int_a^b f(x)\, dx$$

should represent the "area under the curve" $y = f(x)$ for $a \leq x \leq b$.

Definition 1   *Let $[a, b]$ be a given (finite) interval. A **partition** π of $[a, b]$ is a finite set of points $\{x_0, x_1, \ldots, x_n\}$ with*

$$a = x_0 < x_1 < \cdots < x_n = b.$$

We use the notation

$$\Delta x_k = x_k - x_{k-1} \quad (k = 1, 2, \ldots, n).$$

*A second partition π' of $[a, b]$ is said to be **finer** than π if π' contains all the points of π. In this case, we write $\pi' > \pi$. (If $\pi' > \pi$ we also write $\pi < \pi'$ and say that π is **coarser** than π'.)*

Definition 2   *Let f be a bounded function defined on* [a, b] *and let* π *be a partition of* [a, b]. *Then we define the* **upper Darboux sum of f relative to** π *as:*

$$U_\pi(f) = \sum_{k=1}^{n} \left\{ \sup_{[x_{k-1}, x_k]} f \right\} \Delta x_k. \tag{5.3}$$

Similarly, the **lower Darboux sum** *of f relative to* π *is defined to be*

$$L_\pi(f) = \sum_{k=1}^{n} \left\{ \inf_{[x_{k-1}, x_k]} f \right\} \Delta x_k. \tag{5.4}$$

In the above definition, and elsewhere in the sequel, we use a convenient shortened notation for the supremum and infimum of $f(x)$ over $x \in A$:

$$\sup_A f = \sup_{x \in A} f(x); \qquad \inf_A f = \inf_{x \in A} f(x).$$

The geometric meaning of these expressions is illustrated in Figure 5.3, in which the shaded areas represent the upper and lower Darboux sums of f relative to the given partition. It should also be clear from the figure that if

$$A = \text{area under the curve } y = f(x), \quad \text{for } a \le x \le b,$$

then

$$L_\pi(f) \le A \le U_\pi(f). \tag{5.5}$$

This statement cannot be proved as a theorem, for the simple reason that *we do not as yet have a definition for the word "area."* In fact, we are going to use the inequalities (5.5) to *define* area.

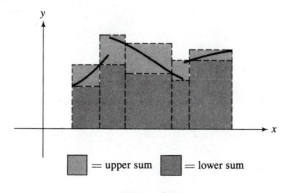

Figure 5.3

Definition 3   *Let f be a bounded function on [a, b]. If there exists a **unique** number A such that (5.5) holds for every partition π of [a, b], we write*

$$A = \int_a^b f(x)\, dx \qquad (5.6)$$

*and call A the **area under the curve** y = f(x) **for** a ≤ x ≤ b. In this case we say that f is **integrable over** [a, b] and A = ∫_a^b f(x) dx is called the (definite) **integral of f over** [a, b].*

We will see shortly that numbers A satisfying (5.5) always exist; it is important to realize that A must be *unique* in order for us to define $\int_a^b f(x)\, dx = A$.

Example 1

Let $f(x) = x$, $0 \le x \le 1$. It is "obvious" that $A = 1/2$. Let us just check that this value $A = 1/2$ is the *only* possible value of A for which (5.5) can hold for every partition π (Figure 5.4).

Suppose $\pi = \left\{ 0, \dfrac{1}{n}, \dfrac{2}{n}, \ldots, \dfrac{n}{n} = 1 \right\}$. Then $\Delta x_k = 1/n$. Also

$$\sup_{[x_{k-1}, x_k]} f = f(x_k) = k/n.$$

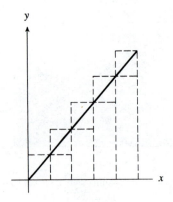

Figure 5.4

Substituting these values in (5.3), we find that

$$U_\pi(f) = \sum_{k=1}^{n} \frac{k}{n^2} = \frac{n+1}{2n} \ .$$

Similarly,

$$L_\pi(f) = \sum_{k=1}^{n} \frac{k-1}{n^2} = \frac{n-1}{2n} \ .$$

Hence, by (5.5), A must satisfy

$$\frac{n-1}{2n} \le A \le \frac{n+1}{2n} \ .$$

Since n is arbitrary, it clearly follows that $A = 1/2$.

Next we give an example of a function f which is so "bizarre" that the concept of "area under f" loses meaning.

Example 2

Let

$$f(x) = \begin{cases} 1 & \text{if } x \text{ is rational,} \\ 0 & \text{if } x \text{ is irrational,} \end{cases}$$

and consider the interval [0, 1]. If $0 \le x_{k-1} < x_k \le 1$, then

$$\sup_{[x_{k-1}, x_k]} f = 1 \quad \text{and} \quad \inf_{[x_{k-1}, x_k]} f = 0.$$

Therefore, for any partition π we have

$$U_\pi(f) = \sum_{k=1}^{n} 1 \cdot (x_k - x_{k-1}) = 1,$$

$$L_\pi(f) = \sum_{k=1}^{n} 0 \cdot (x_k - x_{k-1}) = 0.$$

Thus the inequalities (5.5) are satisfied for any number A with $0 \le A \le 1$. This means that f is *not integrable* over [0, 1].

Recall that the function of Example 2 is everywhere discontinuous (Section 3.4, Example 7). It is known that no such function can be integrated by the Riemann method.†

We come now to the elementary theory of the Riemann integral. Henceforth, let f denote a given bounded function on $[a, b]$.

Lemma   *Let π_1, π_2 be partitions of $[a, b]$. Then*

(i) $L_{\pi_1}(f) \leq U_{\pi_2}(f)$;
(ii)   *if $\pi_1 < \pi_2$, then*

$$L_{\pi_1}(f) \leq L_{\pi_2}(f) \quad and \quad U_{\pi_1}(f) \geq U_{\pi_2}(f).$$

Proof   We prove (ii) first. Suppose to begin with that π_2 contains just one more point x' than π_1, and let $x_{k-1} < x' < x_k$, where x_{k-1}, x_k are points of π_1 (see Figure 5.5). Since obviously $\sup_{[x_{k-1}, x']} f \leq \sup_{[x_{k-1}, x_k]} f$, etc., we have

$$\left\{ \sup_{[x_{k-1}, x']} f \right\}(x' - x_{k-1}) + \left\{ \sup_{[x', x_k]} f \right\}(x_k - x')$$

$$\leq \left\{ \sup_{[x_{k-1}, x_k]} f \right\}(x' - x_{k-1} + x_k - x')$$

$$= \left\{ \sup_{[x_{k-1}, x_k]} f \right\}\Delta x_k.$$

Since the other terms making up $U_{\pi_1}(f)$ and $U_{\pi_2}(f)$ are identical, it follows that $U_{\pi_2}(f) \leq U_{\pi_1}(f)$.

In case π_2 contains several points more than π_1, the desired inequality follows by repeated application of the above argument (i.e. by induction). Similarly we see that $L_{\pi_1}(f) \leq L_{\pi_2}(f)$.

To prove (i) we consider the partition

$$\pi = \pi_1 \cup \pi_2,$$

which consists of all the points of π_1 and of π_2, in order. Then, since $\pi_1 < \pi$ and $\pi_2 < \pi$, it follows from (ii) and the obvious inequality $L_\pi(f) \leq U_\pi(f)$ that

$$L_{\pi_1}(f) \leq L_\pi(f) \leq U_\pi(f) \leq U_{\pi_2}(f). \quad \blacksquare$$

---

† Cf. R. C. James, *Advanced Calculus*, Wadsworth (1966), p. 150.

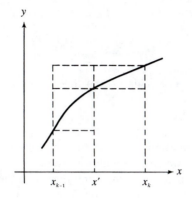

Figure 5.5

Theorem 1   *Let*

$$A_-(f) = \sup_\pi L_\pi(f) \quad \text{and} \quad A_+(f) = \inf_\pi U_\pi(f), \tag{5.7}$$

where the supremum and infimum are taken over all possible partitions π of $[a, b]$. Then

(i) $A_-(f) \le A_+(f)$ *for every function f;*
(ii) $A_-(f) = A_+(f)$ *if and only if f is integrable over $[a, b]$.*

In the latter case we have

$$A_-(f) = A_+(f) = A(f) = \int_a^b f(x)\, dx.$$

Proof   Since $L_{\pi_1}(f) \le U_{\pi_2}(f)$ for all partitions π_1, π_2, it follows that

$$L_{\pi_1} \le \inf_{\pi_2} U_{\pi_2}(f) = A_+(f)$$

and also that

$$A_-(f) = \sup_{\pi_1} L_{\pi_1}(f) \le A_+(f).$$

Hence, for arbitrary partitions π_1, π_2 we have

$$L_{\pi_1}(f) \le A_-(f) \le A_+(f) \le U_{\pi_2}(f).$$

If $A_-(f) \ne A_+(f)$, then (5.5) is satisfied for more than one value of A, and this means that f is not integrable over $[a, b]$.

Conversely suppose $A_-(f) = A_+(f)$, and let A satisfy (5.5). Then

$$A_-(f) = \sup_{\pi_1} L_{\pi_1}(f) \leq A \leq \inf_{\pi_2} U_{\pi_2}(f) = A_+(f),$$

so that $A = A_-(f) = A_+(f)$. This proves that f is integrable over $[a, b]$ and $A = \int_a^b f(x)\,dx = A_-(f) = A_+(f)$.   ∎

Let us survey the situation so far. If we are trying to determine $\int_a^b f(x)\,dx$ by using Darboux sums, we have two alternatives: we can "approximate from above" by taking upper Darboux sums, and then take the infimum of all such approximations; or we can "approximate from below" with lower Darboux sums, in which case we take the supremum. The function f is then *integrable*, according to our definition, if and only if these two methods lead to the same value for $\int_a^b f(x)\,dx$.

In the next theorem we obtain another characterization of integrability, which is useful in the ensuing theory.

Before proceeding, we recall that the equation $\sup_{x \in S} f(x) = \alpha$ holds if and only if $f(x) \leq \alpha$ for every $x \in S$ and, given $\varepsilon > 0$, there is some $x \in S$ such that $f(x) > \alpha - \varepsilon$. These properties of the supremum (and the corresponding properties for the infimum) are used frequently in what follows.

Theorem 2   *The given function f is integrable over $[a, b]$ if and only if, for any given $\varepsilon > 0$, there exists a partition π of $[a, b]$ such that*

$$U_\pi(f) - L_\pi(f) < \varepsilon. \tag{5.8}$$

Proof Suppose f is integrable over $[a, b]$. Then $A_-(f) = A_+(f)$. If $\varepsilon > 0$ is given, let π_1 be a partition such that

$$A_-(f) - L_{\pi_1}(f) < \frac{\varepsilon}{2}$$

and let π_2 be a partition such that

$$U_{\pi_2}(f) - A_+(f) < \frac{\varepsilon}{2}.$$

Let $\pi = \pi_1 \cup \pi_2$; then

$$U_\pi(f) - L_\pi(f) \leq U_{\pi_2}(f) - L_{\pi_1}(f)$$

$$< \frac{\varepsilon}{2} + A_+(f) + \frac{\varepsilon}{2} - A_-(f) = \varepsilon.$$

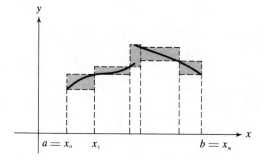

Figure 5.6 $U_\pi(f) - L_\pi(f) = $ (shaded area).

Conversely if (5.8) holds for some partition π, then

$$A_-(f) = \sup_{\pi_1} L_{\pi_1}(f) \geq L_\pi(f)$$

$$> U_\pi(f) - \varepsilon$$

$$\geq A_+(f) - \varepsilon.$$

Since ε is arbitrary, we conclude that $A_-(f) \geq A_+(f)$, and therefore (Figure 5.6) $A_-(f) = A_+(f)$, so that f is integrable over $[a, b]$. ∎

Theorem 2 states that a function is integrable (according to the Riemann-Darboux definition) if and only if the upper approximations $U_\pi(f)$ and the lower approximations $L_\pi(f)$ can be made arbitrarily close to each other by choosing π suitably.

Corollary 1   *If f is monotonic on $[a, b]$, then it is integrable on $[a, b]$.*

Proof   Let π be the "uniform" partition of $[a, b]$, in which

$$x_k - x_{k-1} = \frac{b - a}{n} \quad (k = 1, 2, \ldots, n).$$

Assuming that f is nondecreasing, for example, we have

$$\sup_{[x_{k-1}, x_k]} f = f(x_k) \quad \text{and} \quad \inf_{[x_{k-1}, x_k]} f = f(x_{k-1}).$$

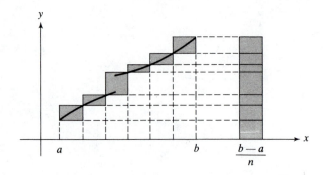

Figure 5.7 $U_\pi(f) - L_\pi(f)$ for increasing f.

Therefore

$$U_\pi(f) - L_\pi(f) = \sum_{k=1}^{n} [f(x_k) - f(x_{k-1})](x_k - x_{k-1})$$

$$= \frac{b-a}{n} \sum_{k=1}^{n} [f(x_k) - f(x_{k-1})]$$

$$= \frac{b-a}{n} [f(b) - f(a)].$$

(See Figure 5.7.) By choosing n large we can therefore make $U_\pi(f) - L_\pi(f)$ arbitrarily small. Hence f is integrable. ∎

Corollary 2   *If f is continuous on $[a, b]$, then f is integrable on $[a, b]$.*

Proof   Let $\varepsilon > 0$ be given. Since f is *uniformly* continuous on $[a, b]$, there exists $\delta > 0$ such that

$$|x - y| < \delta \quad \text{implies} \quad |f(x) - f(y)| < \frac{\varepsilon}{b-a}.$$

Let π be any partition of $[a, b]$ such that $x_k - x_{k-1} < \delta$ $(k = 1, 2, \ldots, n)$. Then

$$\sup_{[x_{k-1}, x_k]} f - \inf_{[x_{k-1}, x_k]} f = \max_{[x_{k-1}, x_k]} f - \min_{[x_{k-1}, x_k]} f < \frac{\varepsilon}{b-a},$$

so that

$$U_\pi(f) - L_\pi(f) < \frac{\varepsilon}{b-a} \sum_{k=1}^{n} (x_k - x_{k-1}) = \varepsilon.$$

Therefore f is integrable on $[a, b]$. ∎

Example 3

An example of an integrable function which is neither monotonic nor continuous is provided by Dirichlet's function (see Figure 5.8),

$$f(x) = \begin{cases} \dfrac{1}{q} & \text{if } x = \dfrac{p}{q} \text{ (in lowest terms),} \\ 0 & \text{if } x \text{ is irrational.} \end{cases}$$

To show that f is integrable over $[0, 1]$, consider a given $\varepsilon > 0$. Let $Q \geq 2$ be an integer with $1/Q < \varepsilon/2$ and let $y_1, y_2, \ldots, y_n$ denote all rational numbers in $[0, 1]$ having denominators $<Q$. Surround each point y_i by an interval I_i of length $\leq \varepsilon/2n$, and let π be the partition consisting of the end points of all these intervals I_i. The intervals of π then consist of (a) the intervals I_i (which can be assumed nonoverlapping), and (b) complementary intervals J_j. For the first we have

$$\sum_i \left(\sup_{I_i} f \right) \cdot (\text{length of } I_i) \leq \sum_{i=1}^{n} 1 \cdot \frac{\varepsilon}{2n} = \frac{\varepsilon}{2}.$$

For the remaining intervals J_j we have $f(x) \leq 1/Q < \varepsilon/2$ for x in J_j, so that

$$\sum_j \left(\sup_{J_j} f \right) \cdot (\text{length of } J_j) < \frac{\varepsilon}{2} \sum_j (\text{length of } J_j) \leq \frac{\varepsilon}{2}.$$

Therefore $U_\pi(f) < \varepsilon$. On the other hand, obviously $L_\pi(f) = 0$; this shows that f is integrable over $[0, 1]$.

The basic properties of the Riemann integral are listed in the following theorem.

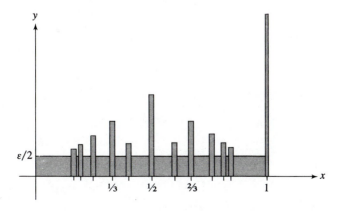

Figure 5.8   Dirichlet's function.

Theorem 3   *Let f and g be integrable functions on [a, b]. Then*

(i) $\displaystyle\int_a^b \alpha f(x)\, dx = \alpha \int_a^b f(x)\, dx$ $(\alpha = \text{constant})$;

(ii) $\displaystyle\int_a^b [f(x) + g(x)]\, dx = \int_a^b f(x)\, dx + \int_a^b g(x)\, dx$;

(iii) $\displaystyle\int_a^b f(x)\, dx = \int_a^c f(x)\, dx + \int_c^b f(x)\, dx$ $(a < c < b)$;

(iv) If $f(x) \le g(x)$ on $[a, b]$,   then $\displaystyle\int_a^b f(x)\, dx \le \int_a^b g(x)\, dx$;

(v) $\displaystyle\left| \int_a^b f(x)\, dx \right| \le \int_a^b |f(x)|\, dx$.

We give here only the proof of part (ii); the remaining parts are outlined in the exercises.

Proof of part (ii)   First we use the fact that (cf. Formula (4.2), page 114)

$$\sup_{x \in S} [f(x) + g(x)] \le \sup_{x \in S} f(x) + \sup_{x \in S} g(x).$$

Therefore,

$$U_\pi(f + g) = \sum_{k=1}^{n} \sup_{x_{k-1} \le x \le x_k} [f(x) + g(x)](x_k - x_{k-1})$$

$$\le U_\pi(f) + U_\pi(g).$$

Similarly,

$$L_\pi(f + g) \ge L_\pi(f) + L_\pi(g).$$

To complete the proof we must use an ε-argument. If $\varepsilon > 0$ is given, there exist partitions π_1 and π_2 of $[a, b]$ such that

$$U_{\pi_1}(f) < \int_a^b f(x)\, dx + \frac{\varepsilon}{2} \quad \text{and} \quad U_{\pi_2}(g) < \int_a^b g(x)\, dx + \frac{\varepsilon}{2}.$$

Let $\pi = \pi_1 \cup \pi_2$; then

$$U_\pi(f + g) \le U_\pi(f) + U_\pi(g)$$

$$\le U_{\pi_1}(f) + U_{\pi_2}(g)$$

$$\le \int_a^b f(x)\, dx + \int_a^b g(x) + \varepsilon.$$

Similarly,

$$L_\pi(f + g) \geq \int_a^b f(x)\, dx + \int_a^b g(x)\, dx - \varepsilon,$$

where π can be assumed (by refinement if necessary) to be the same partition as above. Since we have shown that $U_\pi(f + g) - L_\pi(f + g) < 2\varepsilon$, we conclude that $\int_a^b [f(x) + g(x)]\, dx$ exists. Furthermore,

$$\int_a^b [f(x) + g(x)]\, dx \leq U_\pi(f + g) \leq \int_a^b f(x)\, dx + \int_a^b g(x)\, dx + \varepsilon,$$

for arbitrary ε, and also

$$\int_a^b [f(x) + g(x)]\, dx \geq L_\pi(f + g) \geq \int_a^b f(x)\, dx + \int_a^b g(x)\, dx - \varepsilon.$$

From these inequalities, (ii) is an immediate consequence. ∎

Corollary   *Let f be integrable over $[a, b]$. Then the function*

$$G(x) = \int_a^x f(t)\, dt$$

is continuous on $[a, b]$.

Proof   Let $x_0 < x_1$. Then by Theorem 3

$$|G(x_1) - G(x_0)| = \left| \int_{x_0}^{x_1} f(t)\, dt \right|$$

$$\leq \int_{x_0}^x |f(t)|\, dt$$

$$\leq M(x_1 - x_0),$$

where $M = \sup_{[a, b]} |f(t)|$. Since integrable functions are by definition bounded, M is finite. Continuity of G on $[a, b]$ is therefore proved. ∎

So far we have only defined the integral $\int_a^b f(x)\, dx$ when $a < b$. It is often useful to consider the opposite case.

Definition 4   *If $a > b$, we define*

$$\int_a^b f(x)\, dx = -\int_b^a f(x)\, dx.$$

It is easy to verify that Theorem 3 (suitably modified in some respects) remains valid for the case $a > b$. For the sake of completeness, let us also give Riemann's original definition of the integral as the "limit of a sum." This is the definition usually encountered in calculus texts; it does not require the use of the supremum or infimum. It can be shown by methods like those of this section, that the definitions of Riemann and of Darboux are equivalent.†

Definition 5 (Riemann)   *Let f be a bounded function defined on [a, b]. The Riemann integral of f over [a, b] is defined as follows:*

$$\int_a^b f(x)\, dx = \lim_{\max \Delta x \to 0} \sum_{k=1}^n f(\xi_k)\, \Delta x_k,$$

provided this limit exists, and is independent of the points x_k and ξ_k, $x_{k-1} \leq \xi_k \leq x_k$. Precisely, the limit on the right is defined to equal A if, given any $\varepsilon > 0$, there exists $\delta > 0$ such that for every partition $\pi = \{x_0, x_1, \ldots, x_n\}$ with $\max \Delta x_k < \delta$, and for any points $\xi_1, \xi_2, \ldots, \xi_n$ with $x_{k-1} \leq \xi_k \leq x_k$, we have

$$\left| \sum_{k=1}^n f(\xi_k)\, \Delta x_k - A \right| < \varepsilon.$$

Exercises

1. Let $f(x) = x^2$, $x \in [0, 1]$. If π is the uniform partition $0 < \frac{1}{n} < \frac{2}{n} < \cdots < 1$, calculate $U_\pi(f)$ and $L_\pi(f)$. Find the limits of $U_\pi(f)$ and $L_\pi(f)$ as $n \to \infty$.

2. Let $f(x) = [x]$ (the greatest-integer function. Given $\varepsilon > 0$, find a partition π of the interval $[0, 10]$ with the property that $U_\pi(f) - L_\pi(f) < \varepsilon$.

3. Let $a = x_0 < x_1 < \cdots < x_n = b$ and consider a function $f(x)$ which has a constant value c_k on each interval (x_{k-1}, x_k). (Such a function is called a *step-function*.) Find $\int_a^b f(x)\, dx$.

4. Discuss the role of the symbol "x" in the expression $\int_a^b f(x)\, dx$.

5. (a) Let $f(x) = \sin \frac{1}{x}$ ($x \neq 0$) and $f(0) = 0$. Show that f is integrable on

    $[0, 1]$. (*Hint: f* is certainly integrable over $[\varepsilon, 1]$ for any $\varepsilon > 0$.)

    (b) Generalize the example of part (a) to the case in which f is continuous on $(a, b]$ and bounded on $[a, b]$.

---

† See W. Rudin, *Principles of Mathematical Analysis*, 2d ed., McGraw-Hill (1964), Chapter 6.

6. Let f be a bounded function having a finite number of discontinuities on $[a, b]$. Show that f is integrable on $[a, b]$. (*Hint*: Surround the discontinuities of f by intervals having total length $< \varepsilon/2$. Then use the fact that f is continuous, hence integrable, on the complementary intervals.)

7. Fill in the details of the following proof of Theorem 3(i).

    (a) Suppose $\alpha > 0$. Then $U_\pi(\alpha f) = \alpha U_\pi(f)$, and so on. This implies

    $$A_+(\alpha f) = A_-(\alpha f) = \alpha \int_a^b f(x)\, dx.$$

    (b) Suppose $\alpha = -1$. Then $U_\pi(-f) = -L_\pi(f)$, and so on. Consequently,

    $$A_+(-f) = A_-(-f) = -\int_a^b f(x)\, dx.$$

    (c) Suppose $\alpha = -\beta\,(\beta > 0)$. Then $\int_a^b \alpha f(x)\, dx = \alpha \int_a^b f(x)\, dx$ from parts (a) and (b).

8. Prove Theorem 3(iv); the proof is very simple.

9. Prove Theorem 3(v). (Note that $f(x) \le |f(x)|$.)

10. Complete the details in the proof of the following lemma:

    If f is integrable on $[a, b]$ and $f(x) \equiv 0$ for $x > c$ (where $a < c < b$), then $\int_a^b f(x)\, dx = \int_a^c f(x)\, dx$.

    An outline of the proof is:

    (a) Since f is integrable, for $\varepsilon > 0$ there is a partition π of $[a, b]$ with $c \in \pi$, such that $U_\pi(f) < \int_a^b f(x)\, dx + \varepsilon$. Let π' be the points of π in $[a, c]$. Then $U_\pi(f) = U_{\pi'}(f)$. Consequently (with obvious notation)

    $$A_+(f; [a, c]) < \int_a^b f(x)\, dx + \varepsilon.$$

    Similarly $A_-(f; [a, c]) > \int_a^b f(x)\, dx - \varepsilon$ for arbitrary ε.

    (b) Therefore $A_+(f; [a, c]) = A_-(f; [a, c])$ and the result follows.

11. Derive Theorem 3(iii) from the other parts already proved, by means of the lemma of Exercise 10.

5.3   Integrals and Derivatives

Every calculus student knows that differentiation and integration are inverse processes. For "indefinite" integrals this fact is included in the definition. For "definite" integrals (for example, Riemann integrals) there are two possible formulas:

$$\text{(A)} \quad \frac{d}{dx}\left(\int_a^x f(t)\,dt\right) = f(x);$$

$$\text{(B)} \quad \int_a^x f'(t)\,dt = f(x) - f(a).$$

In this section we prove these formulas under simple continuity hypotheses. Example 2 below indicates that the continuity hypotheses cannot be omitted.

Theorem 1   *Suppose f is defined and continuous on $[a, b]$, and let $x \in (a, b)$. Then*

$$\frac{d}{dx}\left(\int_a^x f(t)\,dt\right) = f(x). \tag{5.9}$$

Proof   From the definition of the derivative, we have

$$\frac{d}{dx}\int_a^x f(t)\,dt = \lim_{h \to 0} \frac{1}{h}\left[\int_a^{x+h} f(t)\,dt - \int_a^x f(t)\,dt\right]$$

$$= \lim_{h \to 0} \frac{1}{h}\int_x^{x+h} f(t)\,dt.$$

For fixed x we can write

$$f(x) = \frac{1}{h}\int_x^{x+h} f(x)\,dt,$$

and therefore

$$\left|\frac{d}{dx}\int_a^x f(t)\,dt - f(x)\right| = \left|\lim_{h \to 0} \frac{1}{h}\int_x^{x+h} [f(t) - f(x)]\,dt\right|$$

$$\leq \lim_{h \to 0} \frac{1}{|h|}\left|\int_x^{x+h} |f(t) - f(x)|\,dt\right|. \tag{5.10}$$

Since f is assumed continuous on $[a, b]$, there exists $\delta > 0$ such that

$$|f(t) - f(x)| < \varepsilon$$

(ε given) when $|t - x| < \delta$. Hence if $|h| < \delta$, we have

$$\frac{1}{|h|}\left|\int_x^{x+h} |f(t) - f(x)|\, dt\right| \leq \frac{1}{|h|} \cdot \varepsilon \cdot |h| = \varepsilon.$$

This shows that the limit on the right side of (5.10) is zero, and (5.9) therefore follows. ∎

Example 1

Given $h(x) = \displaystyle\int_0^{x^2} \frac{\sin t}{t}\, dt$, find $h'(x)$.

Solution We apply Theorem 1 together with the chain rule. Thus let $u = x^2$; then $h(x) = \displaystyle\int_0^u \frac{\sin t}{t}\, dt$. Hence,

$$h'(x) = \frac{dh}{du} \cdot \frac{du}{dx} = \frac{\sin u}{u} \cdot 2x = \frac{2 \sin x^2}{x}.$$

What happens if $f(x)$ is integrable but *not* continuous? The integral

$$F(x) = \int_a^x f(t)\, dt$$

is still defined for all x. By reading through the proof of Theorem 1, the reader will observe that

$$F'(x) = f(x)$$

at *any point of continuity* of $f(x)$. At points of discontinuity of $f(x)$ the derivative of $F(x)$ will generally not exist (although it *may*: see Exercise 10).

Example 2

Consider the "Heaviside" function

$$H(x) = \begin{cases} 0 & \text{if } x < 0 \\ 1 & \text{if } x \geq 0. \end{cases}$$

If $G(x) = \int_0^x H(t)\, dt$ we have

$$G(x) = \begin{cases} 0 & \text{if } x < 0 \\ x & \text{if } x \geq 0 \end{cases}$$

and $G'(x) = H(x)$ except at $x = 0$, where $G'(x)$ does not exist.

Next we consider Formula (B).

Theorem 2 (*The fundamental theorem of calculus*)   *Let f be continuously differentiable on an open interval (α, β). If $\alpha < a < b < \beta$, we have*

$$\int_a^b f'(x)\, dx = f(b) - f(a). \tag{5.11}$$

Proof For $t \in [a, b]$ define

$$G(t) = \int_a^t f'(x)\, dx.$$

From Theorem 1 we see that $G'(t) = f'(t)$ for all t. Therefore $G(t) = f(t) + C$ for some constant C. Setting $t = a$, we obtain $G(a) = 0 = f(a) + C$, so that $C = -f(a)$. Hence $G(t) = f(t) - f(a)$. If we set $t = b$ we obtain (5.11). ∎

The above form of the fundamental theorem of calculus is somewhat unsatisfactory; the basic formula (5.11) refers only to the interval $[a, b]$, but the hypotheses require f to be defined on a larger interval (α, β). By taking slightly more care in the proof, we can overcome this shortcoming, as we have in the following theorem.

Theorem 3 (*Fundamental theorem of calculus: strong form*)   *Let f be continuous on the closed interval $[a, b]$. Let F be defined on the open interval (a, b), and assume that $F'(x) = f(x)$ for $a < x < b$. (F is an "indefinite integral" of f on (a, b)). Then*

$$\int_a^b f(x)\, dx = F(b-) - F(a+), \tag{5.12}$$

where $F(b-) = \lim_{x \to b-} F(x)$ and $F(a+) = \lim_{x \to a+} F(x)$.

Proof For $a \le t \le b$, define

$$G(t) = \int_a^t f(x)\, dx.$$

By Theorem 1 we have $G'(t) = f(t)$ for all t in (a, b). But $f(t) = F'(t)$, and therefore $G(t) = F(t) + C$ ($C = $ const.) for t in (a, b). Letting $t \to a+$, we obtain (since $G(a+) = 0$ by the corollary at the end of the previous section)

$$F(a+) = G(a+) - C = -C.$$

Finally,

$$F(b-) = G(b-) - C$$

$$= \int_a^b f(x)\,dx + F(a+),$$

and this is the same as (5.12).   ∎

Finally, we prove the "integration by parts" formula.

Theorem 4   Let *u* and *v* be continuously differentiable on (α, β). Then for α < a < b < β, we have

$$\int_a^b u(x)v'(x)\,dx = [u(b)v(b) - u(a)v(a)] - \int_a^b u'(x)v(x)\,dx.$$

Proof Let $f(x) = u(x)v(x)$. Apply the fundamental theorem of calculus (Theorem 2) to the function *f*. We leave the calculation to you.   ∎

Exercises

1. Calculate $g'(x)$ if

    (a) $g(x) = \int_0^{2x} e^{-t^2}\,dt,$ (b) $g(x) = \int_x^{2x} e^{-t^2}\,dt.$

2. Show that under the same hypotheses as in Theorem 1,

    $$\frac{d}{dx}\int_x^b f(t)\,dt = -f(x).$$

3. Let $f(x) = [x] + 1$ and define $G(x) = \int_0^x f(t)\,dt$. Determine $G(x)$ explicitly for $0 \le x \le 2$ and verify that $G'(1) \ne f(1)$.

4. Let $g(x) = \int_0^x |t|\,dt$. Does $g'(0)$ exist? Is $g'(x)$ continuous? Differentiable?

5. If *f* is continuous on [a, b] and $G(x) = \int_a^x f(t)\,dt$, show (as in the proof of Theorem 1) that

    $$D^+G(a) = f(a),$$

    where $D^+G(a) = \lim_{x \to a+} \dfrac{G(x) - G(a)}{x - a}$ denotes the *right-hand derivative* of G at $x = a$.

6.   Give an example of a function f which is continuous on $[0, 1]$ and continuously differentiable on $(0, 1)$, but such that f' is unbounded on $(0, 1)$. (Such an example destroys the hope of generalizing the fundamental theorem of calculus too far!)

7.   In elementary calculus, the "indefinite integral," $\int f(x)\, dx$, is defined to be any function $g(x)$ whose derivative equals $f(x)$. Prove from this definition that

(a) $\dfrac{d}{dx} \displaystyle\int f(x)\, dx = f(x),$ (b) $\displaystyle\int f'(x)\, dx = f(x) + c.$

8.   On the basis of Theorem 1 explain why the notion of indefinite integral is unnecessary. Why is this notion nevertheless convenient?

9.   Prove the "change of variables" formula for integrals,

$$\int_a^b f(x)\, dx = \int_\alpha^\beta f(\varphi(t))\varphi'(t)\, dt,$$

making the following assumptions:

(i) f is continuous

(ii) φ is continuously differentiable

(iii) $\varphi(\alpha) = a$ and $\varphi(\beta) = b$.

(*Hint:* Let $F' = f$ and apply the fundamental theorem of calculus.)

*10.   Define $g(x) = \int_0^x \cos(1/t)\, dt$. Show that $g'(0) = 0$, even though $t = 0$ is a point of discontinuity of the integrand. (*Hint:* See Exercise 8 of Section 3.5.)

5.4   Improper Integrals

The definition of the Riemann integral given in Section 5.2 applies only to *bounded functions f* defined on *bounded intervals* $[a, b]$. There are numerous applications, however, in which either the function or the interval becomes unbounded. In such cases we use the concept of an "improper integral."

Example 1

Find $\displaystyle\int_0^1 \dfrac{dx}{\sqrt{x}}$. (See Figure 5.9).

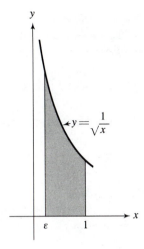

Figure 5.9 $\displaystyle\int_0^1 \frac{dx}{\sqrt{x}}$

Solution Since $1/\sqrt{x}$ is not bounded on $[0, 1]$, this is an "improper integral," which is *defined* to be:

$$\lim_{\varepsilon \to 0+} \int_\varepsilon^1 \frac{dx}{\sqrt{x}} = \lim_{\varepsilon \to 0+} 2(1 - \sqrt{\varepsilon}) = 2.$$

Notice that the area represented by the foregoing integral is the area of an unbounded region. Clearly the only reasonable way to define the area of such a region is to take the limit of the areas of regions "approaching" the given region.

Example 2

$$\int_1^\infty \frac{dx}{\sqrt{x}} = +\infty.$$

This means, by definition, that

$$\lim_{t \to +\infty} \int_1^t \frac{dx}{\sqrt{x}} = +\infty,$$

a fact which can easily be checked.

Definition 1   *Let f be a function defined on (a, b] and such that f is integrable on every interval [a + ε, b] with 0 < ε < b − a. Assume that f is not bounded on (a, b]. We then define the **improper integral** of f over [a, b] as*

$$\int_a^b f(x)\,dx = \lim_{\varepsilon \to 0+} \int_{a+\varepsilon}^b f(x)\,dx,$$

*provided this limit exists. If the limit does exist (finitely) we say that the improper integral **converges**.*

An analogous definition applies to improper integrals of the form $\int_a^\infty f(x)\,dx$. (You should supply this definition.)

Improper integrals have a theory of convergence that is similar in many ways to the theory of infinite series. For example, we have the following "comparison test" for improper integrals.

Theorem 1   *Suppose f and g are defined on [a, ∞) and integrable on [a, b] for every b > a. Suppose also that*

(i) $0 \le f(x) \le g(x)$ *for all x ≥ a,*

(ii) $\int_a^\infty g(x)\,dx$ *converges.*

Then $\int_a^\infty f(x)\,dx$ also converges.

Proof   We can use the simple observation that if h is a nondecreasing function defined on $[a, \infty)$, then $\lim_{x \to \infty} h(x)$ exists (finitely) if and only if h is bounded on $[a, \infty)$.

Now, for $t \ge a$ let

$$F(t) = \int_\alpha^t f(x)\,dx \quad \text{and} \quad G(t) = \int_a^t g(x)\,dx.$$

By hypothesis (ii), $\lim_{t \to \infty} G(t)$ is finite, so that G is bounded on $[a, \infty)$. By hypothesis (i) we have $F(t) \le G(t)$ for all t. Hence F is bounded on $[a, \infty)$ and therefore $\lim_{t \to \infty} F(t)$ is finite.   ∎

The following "improper" form of the fundamental theorem of calculus is easily proved (see Exercise 4):

$$\int_a^\infty F'(x)\,dx = F(\infty) - F(a),$$

where

$$F(\infty) = \lim_{x \to \infty} F(x);$$

this is valid, provided $F'(x)$ is continuous on (α, ∞) for some $\alpha < a$ and the limit $F(\infty)$ exists. There is a similar formula for improper integrals over bounded intervals.

As an application of improper integrals we will consider the famous *Gamma function* $\Gamma(x)$. This function is a generalization of the factorial function $(n!)$.

Definition 2   For $x \geq 1$ define

$$\Gamma(x) = \int_0^\infty t^{x-1} e^{-t}\, dt. \tag{5.13}$$

Theorem 2   *The improper integral (5.13) converges for $x \geq 1$.† Furthermore, we have*

$$\Gamma(x + 1) = x\Gamma(x). \tag{5.14}$$

Proof We know that $\lim_{t\to\infty} t^a e^{-bt} = 0$ for any $a \in \mathbb{R}$ and any $b > 0$. This implies that (with x fixed)

$$t^{x-1} e^{-t} = (t^{x-1} e^{-(1/2)t}) e^{-(1/2)t} < e^{-(1/2)t} \quad \text{for large } t.$$

Since $\int_0^\infty e^{-(1/2)t}\, dt$ obviously converges, the integral (5.13) must also converge, by Theorem 1.

To prove (5.14), we integrate by parts:

$$\Gamma(x + 1) = \int_0^\infty t^x e^{-t}\, dt$$

$$= -t^x e^{-t}\big|_0^\infty + x \int_0^\infty t^{x-1} e^{-t}\, dt$$

$$= x\Gamma(x). \quad \blacksquare$$

Corollary *If n is a positive integer, then $\Gamma(n) = (n-1)!$*

The proof (by induction) is left for you to obtain (see Figure 5.10).

---

† It is easy to show that the integral (5.13) also converges for $0 < x < 1$ (but not for $x \leq 0$). In this case the integral is also improper at $t = 0$.

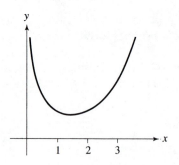

Figure 5.10 $\Gamma(x)$.

Exercises

1. For which values of p does $\int_0^1 x^p\, dx$ converge?

2. For which values of p does $\int_1^\infty x^p\, dx$ converge?

3. Show that the regions whose areas are represented by the following two integrals are congruent:

$$\int_0^1 \left[\frac{1}{\sqrt{x}} - 1\right] dx; \qquad \int_1^\infty \frac{dx}{x^2}.$$

4. (a) Write down the definition of an improper integral of the form $\int_a^\infty f(x)\, dx$.
    (b) Prove the "improper" form of the fundamental theorem of calculus mentioned in the text (page 154).

5. Which of the following improper integrals converge?

    (a) $\displaystyle\int_0^\infty \frac{x^2}{1+x^2}\, dx,$ (b) $\int_0^1 x^{-3/2} \sin x\, dx,$ (c) $\int_{-\infty}^\infty e^{-x^2}\, dx,$

    (d) $\displaystyle\int_e^\infty \frac{dx}{x(\log x)^2},$ (e) $\int_0^1 \log x\, dx.$

6. Calculate

    (a) $\displaystyle\int_{-\infty}^\infty \frac{dx}{1+x^2};$ (b) $\int_0^\infty e^{-x} \sin x\, dx.$

7. Let D denote the 3-dimensional solid interior to the surface obtained by rotating the curve $y = 1/x$ about the x-axis, for $x \geq 1$. Show that D has finite volume but infinite surface area. Is this a paradox?

8. Prove by induction that $\Gamma(n) = (n-1)!$ $(n \geq 1)$.

9. By change of variable in Equation (5.13) show that

$$\Gamma(\tfrac{1}{2}) = 2 \int_0^\infty e^{-y^2} dy.$$

Since the integral on the right is known to equal $\tfrac{1}{2}\sqrt{\pi}$, this shows that

$$\Gamma(\tfrac{1}{2}) = \sqrt{\pi}.$$

10. Using the result of Exercise 9, show by induction that

$$\Gamma(n + \tfrac{1}{2}) = \frac{(2n)!\sqrt{\pi}}{4^n n!} \quad \text{for } n = 0, 1, 2, \ldots.$$

5.5 Uniform Convergence

Let $\{f_n\}$ be a given sequence of functions defined on a fixed interval I. If $\lim_{n \to \infty} f_n(x)$ exists for each point x in I, we say that the function f defined by

$$f(x) = \lim_{n \to \infty} f_n(x) \tag{5.15}$$

is the *pointwise limit* of the sequence $\{f_n\}$ on I.

The concept of pointwise limit, which at first seems to be very natural, suffers from several shortcomings. For example,

(a) It does not follow that the function f is continuous, even if every function f_n is continuous. (See Example 1.)

(b) It does not follow that $f'(x) = \lim_{n \to \infty} f_n'(x)$, even if every function f_n is differentiable. (See Example 2.)

(c) It does not follow that $\int_I f(x)\,dx = \lim_{n \to \infty} \int_I f_n(x)\,dx$, even if this limit exists. (See Example 3.)

Example 1

Let $f_n(x) = \dfrac{x^n}{1 + x^n}$ $(0 \leq x < +\infty)$. It is easy to verify (see Figure 5.11) that

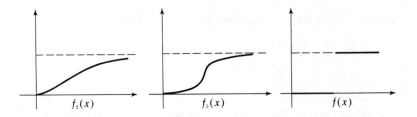

Figure 5.11   The sequence of Example 1.

$$\lim_{n \to \infty} f_n(x) = \begin{cases} 0 & \text{if } 0 \leq x < 1, \\ \frac{1}{2} & \text{if } x = 1, \\ 1 & \text{if } x > 1. \end{cases}$$

Notice that although each function f_n is continuous, the limit function f is discontinuous at $x = 1$.

Example 2

Let $g_n(x) = (\sin nx)/n$. Clearly $\lim_{n \to \infty} g_n(x) = 0$ for every x. However, since $g_n'(x) = \cos nx$, $\lim_{n \to \infty} g_n'(x)$ does not exist except at $x = 2k\pi$, k an integer. (See Figure 5.12.)

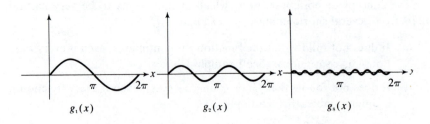

Figure 5.12   The sequence of Example 2.

Example 3

Let $h_n(x) = nx^n$ for $0 < x < 1$. Then $\lim_{n \to \infty} h_n(x) = 0$ for every x in $(0, 1)$. However,

$$\int_0^1 h_n(x)\, dx = n \int_0^1 x^n\, dx = \frac{n}{n+1} \to 1 \quad \text{as } n \to \infty.$$

This shows that (see Figure 5.13)

$$\lim_{n \to \infty} \int_0^1 h_n(x)\, dx \neq \int_0^1 \lim_{n \to \infty} h_n(x)\, dx.$$

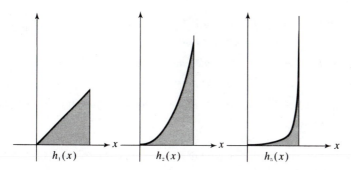

Figure 5.13 The sequence of Example 3. (The areas of the shaded regions $\to 1$.)

You will perhaps raise an objection about Example 3; it certainly seems unreasonable to say that the functions h_n approach the function $h(x) \equiv 0$ on $(0, 1)$. Although it is true that $h_n(x) \to 0$ for each given point $x \in (0, 1)$, the functions $h_n(x)$ with large n do not "look" much like the zero function. To make this more precise, we introduce the following definitions.

Definition 1   If f and g are two *bounded functions* defined on an interval I, we define the **distance between** f and g to be

$$d(f, g) = d_I(f, g) = \sup_{x \in I} |f(x) - g(x)|. \tag{5.16}$$

Definition 2   Let $\{f_n\}$ be a *sequence of functions* defined on I and f another function defined on I. If

$$d_I(f_n, f) \to 0 \quad \text{as} \quad n \to \infty,$$

we say that f_n **converges uniformly on** I to f.

Returning to Example 3, notice that

$$d_{(0,1)}(h_n, 0) = \sup_{0 < x < 1} |nx^n| = n.$$

Therefore h_n does not converge uniformly to zero on $(0, 1)$. The student should check Examples 1 and 2 for uniform convergence; thus for Example 1, one can compute $d(f_n, f)$ and decide whether $d(f_n, f) \to 0$ as $n \to \infty$.

To explain why $d(f, g)$ is called the "distance" between f and g, we list some simple properties of $d(f, g)$:

Lemma 1   *The distance $d(f, g)$ between bounded functions f, g on a given interval I satisfies*

  (i) $d(f, g) = 0$ *if and only if $f(x) \equiv g(x)$ on I,*
  (ii) $d(f, g) \geq 0$ *for all functions f, g,*
  (iii) $d(f, g) = d(g, f)$ *for all functions f, g,*
  (iv) $d(cf, cg) = |c| d(f, g)$ *for all functions f, g and constants c,*
  (v) $d(f, g) \leq d(f, h) + d(h, g)$ (**triangle inequality**) *for all functions f, g, h.*

Proof Properties (i)–(iv) are trivial from (5.16). To prove (v), we have (by Inequality (4.2), page 114)

$$d(f, g) = \sup_{x \in I} |f(x) - g(x)|$$

$$\leq \sup_{x \in I} (|f(x) - h(x)| + |h(x) - g(x)|)$$

$$\leq \sup_{x \in I} |f(x) - h(x)| + \sup_{x \in I} |h(x) - g(x)|$$

$$= d(f, h) + d(h, g). \quad \blacksquare$$

Notice that, graphically speaking, the distance $d(f, g)$ represents the "greatest" (= supremum) vertical distance between points on the graphs of f and g having equal abscissae (Figure 5.14).

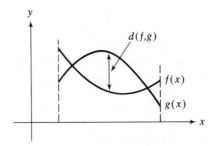

Figure 5.14

The definition of uniform convergence can be given in "ε-δ" terms as shown in the following lemma.

Lemma 2   *The sequence $\{f_n\}$ converges uniformly to f on the interval I if and only if the following condition is satisfied:*

given $\varepsilon > 0$, *there exists an integer N such that, for all $x \in I$, we have*

$$|f_n(x) - f(x)| < \varepsilon \quad whenever \; n \geq N.$$

Proof   Suppose $f_n \to f$ uniformly on I. Then $d(f_n, f) \to 0$. Thus if $\varepsilon > 0$ is given, there exists an integer N such that

$$\sup_{x \in I} |f_n(x) - f(x)| = d_I(f_n, f) < \varepsilon \quad \text{for all } n \geq N.$$

Consequently, for every $x \in I$,

$$|f_n(x) - f(x)| < \varepsilon, \quad \text{if } n \geq N.$$

The converse is proved similarly. ∎

Example 4

Let $f_n(x) = x^n$. We can show that $f_n \to 0$ uniformly on the interval $[-a, a]$, provided $0 < a < 1$, but $\{f_n\}$ does not converge uniformly on $[-1, 1]$. (See Figure 5.15.)

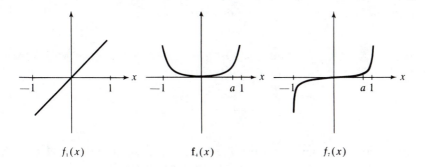

$$f_1(x) \qquad\qquad f_4(x) \qquad\qquad f_7(x)$$

Figure 5.15 $f_n(x) = x^n$ on $[-1, 1]$

For, consider

$$d_{[-a,a]}(f_n, 0) = \sup_{|x| \le a} x^n = a^n$$

$$\to 0 \quad \text{as} \quad n \to \infty \quad \text{if } 0 < a < 1.$$

This shows that $f_n \to 0$ uniformly on $[-a, a]$. Since $f_n(-1) = (-1)^n$, the sequence $\{f_n\}$ does not converge at $x = -1$; that is, the sequence $\{f_n\}$ does not converge pointwise on the interval $[-1, 1]$. (The sequence does converge at $x = 1$; we leave it to you to decide whether $\{f_n\}$ converges uniformly on $[0, 1]$, for example.)

We are prepared now to prove that uniform convergence does not have the shortcomings [$(a) - (c)$ above] that pointwise convergence has.

Theorem 1   *Let $\{f_n\}$ be a sequence of continuous functions on an interval I. If $f_n \to f$ uniformly on I, then f is also continuous on I.*

Proof   Fix $x_0 \in I$, and let $\varepsilon > 0$. By the triangle inequality, for any $x \in I$,

$$|f(x) - f(x_0)| \le |f(x) - f_n(x)| + |f_n(x) - f_n(x_0)| + |f_n(x_0) - f(x_0)|. \quad (5.17)$$

Since $f_n \to f$ uniformly on I, we can find an integer N such that $|f_N(z) - f(z)| < \varepsilon/3$ for all $z \in I$. Since f_N is continuous on I, there exists a $\delta > 0$ such that $|f_N(x) - f_N(x_0)| < \varepsilon/3$ provided $|x - x_0| < \delta$. Therefore, by (5.17), we have

$$|f(x) - f(x_0)| < \varepsilon \quad \text{provided } |x - x_0| < \delta.$$

This proves that f is continuous at x_0. ∎

The proof of Theorem 1 could be summarized as follows: since each f_n is continuous at x_0, and since $f_n(x)$ is "arbitrarily close" to $f(x)$ for all x, it follows that f must be continuous at x_0.

Theorem 2   *Let $\{f_n\}$ be a sequence of continuous functions on a closed, bounded interval $[a, b]$. If $f_n \to f$ uniformly on $[a, b]$, then*

$$\int_a^b f_n(x)\, dx \to \int_a^b f(x)\, dx \quad \text{as } n \to \infty. \quad (5.18)$$

Proof   First, f is continuous on $[a, b]$ according to Theorem 1. Hence, the functions f_n ($n = 1, 2, 3, \ldots$) and f are all integrable over $[a, b]$.

If $\varepsilon > 0$ is given, choose an integer N such that

$$|f_n(x) - f(x)| < \frac{\varepsilon}{b - a} \quad \text{for all } x \in [a, b] \quad \text{and all } n \geq N.$$

Then

$$\left| \int_a^b f_n(x)\, dx - \int_a^b f(x)\, dx \right| \leq \int_a^b |f_n(x) - f(x)|\, dx < \varepsilon$$

for $n \geq N$.   ∎

Theorem 2 can be generalized as follows: if the functions f_n are assumed integrable (but not necessarily continuous) over $[a, b]$, and $f_n \to f$ uniformly on $[a, b]$, then f is also integrable on $[a, b]$, and (5.18) holds. See Exercise 10.

Theorem 3   *Let $\{f_n\}$ be a sequence of continuously differentiable functions on an open interval I. Suppose that for some functions f and g on I we have*

(a) *$f_n \to f$ uniformly on I, and*
(b) *$f_n' \to g$ uniformly on I.*

Then f is continuously differentiable on I, and we have $f' = g$. Thus $f_n' \to f'$ uniformly on I.

Proof   We use Theorem 2 together with some theorems of Section 5.3. Let a be a fixed point of I. Then for any $x \in I$,

$$\int_a^x f_n'(t)\, dt = f_n(x) - f_n(a) \tag{5.19}$$

by the fundamental theorem of calculus.

Letting $n \to \infty$ and applying Theorem 2 to the left side of (5.19), we obtain

$$\int_a^x g(t)\, dt = f(x) - f(a). \tag{5.20}$$

Differentiation of (5.20) yields (by Theorem 1, Section 5.3)

$$g(x) = f'(x).   ∎$$

We note that Theorem 3 becomes false if condition (b) is omitted (Exercise 7).

This completes our study of the *consequences* of uniform convergence. We end this section with a famous and useful criterion that is necessary and sufficient for uniform convergence.

Theorem 4 (Cauchy's criterion for uniform convergence)   Let $\{f_n\}$ be a sequence of functions on an interval I. Then f_n converges uniformly on I to some function f if and only if

$$\lim_{n,m \to \infty} d_I(f_n, f_m) = 0,$$

or, in other words, if and only if, given $\varepsilon > 0$, there exists an integer N such that

$$|f_n(x) - f_m(x)| < \varepsilon \quad \text{for all } x \in I \text{ and all } n, m \geq N. \tag{5.21}$$

Proof   Suppose first that $f_n \to f$ uniformly on I, that is, $d_I(f_n, f) \to 0$ as $n \to \infty$. By Lemma 1 (v) (the triangle inequality), we conclude that

$$d_I(f_n, f_m) \leq d_I(f_n, f) + d_I(f_m, f) \to 0 \quad \text{as} \quad n, m \to \infty.$$

The converse is more difficult to prove. If (5.21) is satisfied, it follows that for each given $x \in I$ the numerical sequence $\{f_n(x)\}$ is a Cauchy sequence. Hence (Section 4.4), $f(x) = \lim_{n\to\infty} f_n(x)$ exists for each x.

We must show that $f_n \to f$ uniformly on I. Let $\varepsilon > 0$ be given. By (5.21) there exists an integer N such that

$$|f_n(x) - f_m(x)| < \frac{\varepsilon}{2} \quad \text{for all } x \in I \quad \text{and all } n, m \geq N.$$

If x is a given point of I, then, since $f_n(x) \to f(x)$, there exists an integer $N_1 \geq N$ such that $|f_{N_1}(x) - f(x)| < \varepsilon/2$.† Hence for $n \geq N$ and for any given $x \in I$, we have

$$|f_n(x) - f(x)| \leq |f_n(x) - f_{N_1}(x)| + |f_{N_1}(x) - f(x)| < \varepsilon.$$

Thus $d_I(f_n, f) \leq \varepsilon$ for $n \geq N$, and this proves that $f_n \to f$ uniformly on I.   ∎

Corollary   Let $\{f_n\}$ be a sequence of continuous functions on an interval I, and suppose $\{f_n\}$ satisfies the "uniform Cauchy condition" (5.21). Then $\{f_n\}$ converges uniformly on I to a continuous function f.

This result follows by combining Theorems 4 and 1.

---

† Note that for all we know at this stage of the proof, N_1 might depend on x (as well as ε), but this does not affect the validity of the proof. Of course once the proof has been completed, we'll know that in fact N_1 could be chosen to be independent of x.

Exercises

1. Consider the intervals

$$I_1 = [0, +\infty); \qquad I_2 = [0, 1]; \qquad I_3 = [1, +\infty).$$

For each of the following sequences of functions,

(i) find the pointwise limit, $\lim_{n\to\infty} f_n(x) = f(x)$;

(ii) determine whether $f_n \to f$ uniformly on each interval I_1, I_2, and I_3.

(a) $f_n(x) = \dfrac{x}{n}$,   (b) $f_n(x) = x^{1/n}$,   (c) $f_n(x) = e^{-nx}$.

2. Consider $f_n(x) = nx^n(1 - x)$ on $[0, 1]$.

  (a) Find $f(x) = \lim_{n\to\infty} f_n(x)$.

  (b) Calculate $d_{[0,1]}(f_n, f)$ by calculus.

  (c) Does $f_n \to f$ uniformly on $[0, 1]$?

  (d) Does $\int_0^1 f_n(x)\, dx \to \int_0^1 f(x)\, dx$?

3. Repeat Exercise 2 for the sequence $g_n(x) = n^2 x^n(1 - x)$.

4. What does the statement "the infinite series $\sum_1^\infty f_n$ converges uniformly to f on the interval I" mean precisely?

5. Show that the series $\sum_0^\infty x^n$ converges uniformly on $[-a, a]$ to $\dfrac{1}{1 - x}$ provided $0 < a < 1$. (See Example 4.)

6. If $f(x)$ is defined on I, then the set of all points (x, y) in the plane satisfying

$$|y - f(x)| < \varepsilon, \quad x \in I,$$

is called the *ε-strip* about the curve $y = f(x)$.

  (a) Sketch the 0.2-strip of the curve $y = x^3$ on $[0, 1]$.

  (b) Express the concept of uniform convergence in terms of ε-strips, and draw a picture.

7. Show by example that

$$f_n \to 0 \text{ uniformly} \quad \text{and} \quad f_n' \text{ exists},$$

do not imply $f_n' \to 0$.

8. Prove that if $\{f_n\}$ is a sequence of functions which are uniformly continuous on an interval I and if $f_n \to f$ uniformly on I, then f is also uniformly continuous on I.

9. Let $\{f_n\}$ be a sequence of nondecreasing functions on $[a, b]$, and suppose $f_n \to 0$ pointwise on $[a, b]$. Show that $f_n \to 0$ uniformly on $[a, b]$.

*10. Prove that if $\{f_n\}$ is a sequence of integrable functions on $[a, b]$ and if $f_n \to f$ uniformly on $[a, b]$, then f is also integrable on $[a, b]$ and

$$\int_a^b f_n(x)\,dx \to \int_a^b f(x)\,dx.$$

(*Hint:* You wish to show that, given $\varepsilon > 0$, there exists a partition π of $[a, b]$ such that $U_\pi(f) - L_\pi(f) < \varepsilon$.)

11. Define

$$f_n(x) = \begin{cases} 1 & \text{if } \dfrac{1}{2n} < x < \dfrac{1}{n} \\ 0 & \text{for all other } x \text{ in } [0, 1]. \end{cases}$$

Show that $f_n \to 0$ pointwise, but not uniformly on $[0, 1]$. Does

$$\lim_{n \to \infty} \int_0^1 f_n(x)\,dx = \int_0^1 \lim_{n \to \infty} f_n(x)\,dx?$$

12. Define

$$f_n(x) = \begin{cases} n & \text{if } \dfrac{1}{2n} < x < \dfrac{1}{n} \\ 0 & \text{for all other } x \text{ in } [0, 1]. \end{cases}$$

Answer the same questions as in Exercise 11.

5.6   Power Series

We recall from Sec. 2.4 that a series of the form

$$\sum_{n=0}^{\infty} a_n(x - x_0)^n \tag{5.22}$$

is called a *power series in x*, with *center x_0* and *coefficients $a_0, a_1, a_2, \ldots$*. Since there is no loss of generality in the theory if we assume that $x_0 = 0$, we will mainly consider power series with center 0:

$$\sum_{n=0}^{\infty} a_n x^n. \tag{5.23}$$

In Sec. 2.4 we prove that the series (5.23) had an "interval of convergence" $|x| < R = 1/\alpha$, *provided* that the limit

$$\alpha = \lim_{n \to \infty} \left| \frac{a_{n+1}}{a_n} \right| \tag{5.24}$$

existed. In this section we will prove that *every* power series has such an interval of convergence, and we will give a different formula for α, which is valid under all circumstances.

Example 1

Let

$$a_n = \begin{cases} 1 & \text{if } n \text{ is even} \\ 1/n & \text{if } n \text{ is odd.} \end{cases}$$

It is easy to see that the limit (5.24) does not exist in this case, yet the power series $\sum_0^{\infty} a_n x^n$ can easily be shown to have radius of convergence $R = 1$.

We introduce the following notion of the "limit superior" of a sequence in order to obtain the desired results. This notion arises in many other applications.

Definition   *Let $\{b_n\}$ be a sequence of real numbers. We write*

$$\limsup_{n \to \infty} b_n = \beta,$$

provided that the following two conditions are satisfied:

(1) Given any $\varepsilon > 0$, there exists an integer N such that

$$b_n < \beta + \varepsilon \quad \text{for all } n \geq N,$$

(2) Given any $\varepsilon > 0$ and any integer i, there exists an integer $j \geq i$ such that $b_j > \beta - \varepsilon$.

If the sequence $\{b_n\}$ is not bounded from above, we define

$$\limsup_{n\to\infty} b_n = +\infty;$$

finally if $\lim_{n\to\infty} b_n = -\infty$, we define

$$\limsup_{n\to\infty} b_n = -\infty.$$

Example 2

Let $b_n = (-1)^n + 1/n$. Then

$$\limsup_{n\to\infty} b_n = 1.$$

To verify this we merely have to check that conditions (1) and (2) hold. However, these are both obvious.

Note that the sequence $\{b_n\}$ of this example does not converge, but it has a subsequence, $\{b_{2n}\}$, converging to 1. In general we have: *for any sequence $\{b_n\}$, $\limsup_n b_n$ is the largest number that can be obtained as a limit of some subsequence of $\{b_n\}$.* This is easy to see from the Definition: Condition (1) implies that no subsequence can converge to a number $> \beta = \limsup_{n\to\infty} b_n$; Condition (2) says that some subsequence does converge to β.

Theorem 1   *If $\{b_n\}$ is any given sequence, then*

$$\beta = \limsup_{n\to\infty} b_n$$

exists $(-\infty \leq \beta \leq +\infty)$ and is uniquely determined.

The proof is easy and will be omitted.

Theorem 2 (J. Hadamard)   *Let $\sum_{n=0}^{\infty} a_n x^n$ be a given power series. Define*

$$\alpha = \limsup_{n\to\infty} |a_n|^{1/n}. \tag{5.25}$$

Let $R = 1/\alpha$ *(if* $\alpha = 0$, *let* $R = +\infty$; *if* $\alpha = +\infty$, *let* $R = 0$). *Then the series* $\sum_{n=0}^{\infty} a_n x^n$

(a) *converges absolutely for all* $|x| < R$,

(b) *diverges for all* $|x| > R$.

This theorem generalizes the formula (2.21) for the reciprocal of the radius of convergence R of a power series, and shows that *every* power series has a radius of convergence.

Proof   (a) Suppose first that $\alpha \neq 0$, $+\infty$. If $|x| < R = 1/\alpha$, then for some $\varepsilon > 0$ we have $|x| < 1/\alpha(1 + \varepsilon)$. By Equation (5.25) we have $|a_n|^{1/n} < \alpha(1 + \varepsilon/2)$ for sufficiently large n. Hence for such n,

$$|a_n x^n| < \left(\frac{1 + \varepsilon/2}{1 + \varepsilon}\right)^n = \tau^n, \tag{5.26}$$

where $\tau < 1$. Thus the series $\sum_{n=0}^{\infty} a_n x^n$ converges absolutely by comparison with the geometric series $\sum_0^{\infty} \tau^n$.

In the case $\alpha = 0$ we argue as follows. Given $x \neq 0$, by the definition we have $|a_n|^{1/n} < 1/2 |x|$ for large n. Therefore $|a_n x^n| < 1/2^n$ for large n, so that $\sum_0^{\infty} a_n x^n$ converges absolutely. The case $\alpha = \infty$ is trivial.

(b) Suppose that $\sum_{n=0}^{\infty} a_n x^n$ converges for a particular $x \neq 0$. Then $a_n x^n \to 0$ as $n \to \infty$, and therefore

$$|a_n x^n|^{1/n} = |a_n|^{1/n} |x| < 1$$

for large n. Thus

$$\alpha = \limsup_{n \to \infty} |a_n|^{1/n} \leq \frac{1}{|x|},$$

so that $|x| \leq 1/\alpha = R$. This proves (b). ∎

For the power series of Example 1 above we clearly have $R = 1$, since in this case

$$\alpha = \limsup_{n \to \infty} |a_n|^{1/n} = \lim_{n \to \infty} |a_n|^{1/n} = 1.$$

Other examples are given in the exercises.

Exercises

1. Find by inspection the limit superior of each of the following sequences:

    (a) $1, \dfrac{1}{2}, \dfrac{3}{2}, \dfrac{1}{4}, \dfrac{7}{4}, \ldots, \dfrac{1}{2^n}, 2 - \dfrac{1}{2^n}, \ldots$;

    (b) $(-1)^n$;

    (c) $1, 2, \tfrac{1}{2}, 4, \tfrac{1}{4}, 8, \ldots$;

    (d) $\left| (-1)^n + \dfrac{n}{n+1} \right|$;

    (e) $\dfrac{n}{3} - \left[\dfrac{n}{3} \right]$ ([ ] $=$ greatest integer).

2. Use Hadamard's formula (5.25) to find the radii of convergence of the following power series.

    (a) $\displaystyle\sum_{0}^{\infty} n^2 x^n$ (b) $\displaystyle\sum_{0}^{\infty} 2^{n+1} x^n$

    (c) $\displaystyle\sum_{0}^{\infty} 2^{n(-1)^n} x^n$ (d) $\displaystyle\sum_{0}^{\infty} n x^{n^2}$

    For which of these series does the limit (5.24) exist?

3. (a) Define $\displaystyle\liminf_{n \to \infty} b_n$ and discuss it.

    (b) Find the limit inferior of the sequences of Exercise 1.

4. Show that if $c > 0$, then

$$\limsup_{n \to \infty} c b_n = c \limsup_{n \to \infty} b_n.$$

    What if $c < 0$?

5. Show that, in general,

$$\limsup_{n \to \infty} (x_n + y_n) \neq \limsup_{n \to \infty} x_n + \limsup_{n \to \infty} y_n.$$

    What *can* be said in this regard?

6. If $\{b_n\}$ converges, show that

$$\limsup_{n \to \infty} b_n = \lim_{n \to \infty} b_n.$$

7. If the coefficients in a power series are integers, and if the series is not a polynomial, show that $R \le 1$.

8. Let $\{b_n\}$ be a bounded sequence. Prove that

$$\limsup_{n \to \infty} b_n = \inf_{n \ge 1} (\sup_{k \ge n} b_k).$$

What is the corresponding formula for $\liminf_{n \to \infty} b_n$?

9. Show that if $a_n \to 0$ as $n \to \infty$ then $\sum_0^\infty a_n x^n$ converges for $|x| < 1$.

10. What can you say about the radius of convergence of the series

$$\sum_0^\infty (a_n + b_n)x^n$$

in terms of the radii of convergence of the two series $\sum_0^\infty a_n x^n$ and $\sum_0^\infty b_n x^n$?

11. For what values of x does the series

$$\sum_{n=0}^\infty a_n x^{-n}$$

converge?

12. Define convergence for *double infinite series*:

$$\sum_{n=-\infty}^\infty a_n.$$

(You may wish to adapt your definition to the following exercise.)

13. Show that a series of the form

$$\sum_{n=-\infty}^\infty a_n x^n$$

has an "annulus of convergence" $r < |x| < R$ (which may be empty). Give formulas for r, R, and give examples. (See Exercises 11 and 12.)

5.7   Uniform Convergence of Power Series

We prove next the following fundamental result.

Theorem 1   Let $\sum_0^\infty a_n x^n$ be a given power series with radius of convergence R. If r is any number such that $0 < r < R$, then the given series converges uniformly on the interval $[-r, r]$.

Proof   Suppose first that $\alpha \neq 0$, where α is given by (5.25). Since $r < R = 1/\alpha$, we can write

$$r = \frac{1}{\alpha(1 + \varepsilon)} \quad \text{(some } \varepsilon > 0\text{)}.$$

By (5.26), if $|x| \leq r$, we have

$$|a_n x^n| \leq \left(\frac{1 + \varepsilon/2}{1 + \varepsilon}\right)^n = \tau^n \quad \text{for large } n,$$

where $\tau < 1$.

We can now apply the uniform Cauchy criterion (Theorem 4, Section 5.5); for $m \geq n$ we have (if $|x| \leq r$)

$$\left| \sum_{k=n}^{m} a_k x^k \right| \leq \sum_{k=n}^{m} \tau^k \to 0 \quad \text{as} \quad n, m \to +\infty,$$

since the geometric series $\sum_1^\infty \tau^k$ converges. Therefore, the series $\sum_0^\infty a_k x^k$ converges uniformly on the interval $|x| \leq r$.

If $\alpha = 0$, let $r > 0$ be given. Then $|a_n|^{1/n} < 1/2r$ for large n, so that

$$|a_n x^n| < 1/2^n \quad \text{for } |x| \leq r \quad \text{and} \quad n \text{ large}.$$

Hence $\sum_0^\infty a_n x^n$ converges uniformly for $|x| \leq r$. ∎

Corollary   The power series $\sum_0^\infty a_n x^n$ converges to a continuous function on the open interval $(-R, R)$.

Proof   The partial sums $\sum_0^N a_n x^n$, being polynomials, are continuous for every x. Since these partial sums converge uniformly to $f(x) = \sum_0^\infty a_n x^n$ on any interval $[-r, r]$ $(r < R)$, the sum $f(x)$ is also continuous on $[-r, r]$ for every $r < R$. Hence, $f(x)$ is continuous on $(-R, R)$. ∎

Example 1

The geometric series $\sum_0^\infty x^n$ converges uniformly on $[-r, r]$ $(r < 1)$ to $f(x) = 1/(1 - x)$; the sum is continuous on $(-1, 1)$ but certainly not on $[-1, 1]$. (See Figure 5.16.)

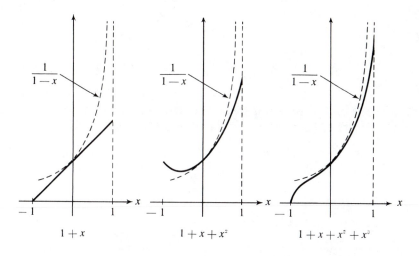

$$1 + x \qquad\qquad 1 + x + x^2 \qquad\qquad 1 + x + x^2 + x^3$$

Figure 5.16 Partial sums of the geometric series $\sum_0^\infty x^n$.

Theorem 2 *Let* $f(x) = \sum_0^\infty a_n x^n$ *have radius of convergence R. Then*

$$\int_0^x f(t)\, dt = \sum_0^\infty \frac{a_n}{n + 1} x^{n+1} \quad (|x| < R) \tag{5.27}$$

and the power series on the right side also has radius of convergence R.

Proof Let R' denote the radius of convergence of the power series (5.27):

$$\frac{1}{R'} = \limsup_{n \to \infty} \left| \frac{a_n}{n + 1} \right|^{1/n} = \limsup_{n \to \infty} \left\{ |a_n|^{1/n} \cdot \frac{1}{(n + 1)^{1/n}} \right\}$$

Since $\lim_{n \to \infty} (n + 1)^{1/n} = 1$, it follows from the Lemma below that

$$\frac{1}{R'} = \limsup_{n \to \infty} |a_n|^{1/n} = \frac{1}{R}.$$

Now if $0 < |x| < R$, the series $\sum_0^\infty a_n t^n$ converges uniformly to $f(t)$ on the interval between 0 and x. By Theorem 2, Section 5.5, this implies that

$$\lim_{n \to \infty} \int_0^x \sum_{k=0}^n a_k t^k \, dt = \int_0^x f(t) \, dt.$$

But

$$\lim_{n \to \infty} \int_0^x \sum_{k=0}^n a_k t^k \, dt = \lim_{n \to \infty} \sum_{k=0}^n a_k \int_0^x t^k \, dt$$

$$= \lim_{n \to \infty} \sum_{k=0}^n \frac{a_k}{k+1} x^{k+1}$$

$$= \sum_0^\infty \frac{a_n}{n+1} x^{n+1}. \quad \blacksquare$$

The result just proved is frequently stated in the form: "any power series can be integrated term by term within its interval of convergence."

Example 2

Integrating the geometric series

$$\sum_0^\infty x^n = \frac{1}{1-x} \quad (|x| < 1)$$

term by term, we obtain

$$\sum_0^\infty \frac{x^{n+1}}{n+1} = \int_0^x \frac{dt}{1-t} = \log(1-x) \quad (|x| < 1).$$

Replacing x by $-x$, we obtain the formula

$$\log(1+x) = \sum_{n=1}^\infty \frac{(-1)^{n+1} x^n}{n}$$

$$= x - \frac{x^2}{2} + \frac{x^3}{3} - \cdots \quad (|x| < 1). \qquad (5.28)$$

We need to prove the following lemma, which generalizes Exercise 4 of the preceding section.

Lemma   *Suppose that* $\lim\limits_{n\to\infty} x_n = \alpha > 0$ *and* $\limsup\limits_{n\to\infty} y_n = \beta$. *Then*

$$\limsup_{n\to\infty} x_n y_n = \alpha\beta. \qquad (5.29)$$

Proof   We recall the characterization (page 168) of $\limsup_{n\to\infty} b_n$ as the largest number that can be obtained as the limit of some subsequence of $\{b_n\}$. Thus we have $y_{n_k} \to \beta$ as $k \to \infty$ for some subsequence of $\{y_n\}$. Therefore $x_{n_k} y_{n_k} \to \alpha\beta$ as $k \to \infty$, i.e. $\alpha\beta$ is obtained as the limit of a subseqence of $\{x_n y_n\}$.

Next, for any subsequence $\{x_{n'_k} y_{n'_k}\}$ converging, say to γ, we have $y_{n'_k} \to \gamma/\alpha$ as $k \to \infty$, so that $\gamma/\alpha \le \beta$ by the above characterization. Thus $\gamma \le \alpha\beta$, which tells us that $\alpha\beta$ is the largest number obtainable as the limit of some subsequence of $\{x_n y_n\}$. Hence $\limsup_{n\to\infty} x_n y_n = \alpha\beta$. ∎

We turn next to the question of differentiation of power series.

Theorem 3   *Let* $f(x) = \sum_0^\infty a_n x^n$ *have radius of convergence R. Then*

$$f'(x) = \sum_0^\infty n a_n x^{n-1} \quad (|x| < R), \qquad (5.30)$$

and this power series also has radius of convergence R.

Proof   That the series (5.30) has radius of convergence R follows exactly as in the previous theorem; we omit the details. Let

$$g(x) = \sum_0^\infty n a_n x^{n-1} \quad (|x| < R).$$

By Theorem 2 we have

$$\int_0^x g(t)\, dt = \sum_0^\infty a_n x^n = f(x) \quad (|x| < R).$$

Hence, by differentiation, $g(x) = f'(x)$. ∎

Corollary   *Let* $f(x) = \sum_0^\infty a_n x^n$ *have radius of convergence R. Then f is infinitely differentiable on the open interval* $|x| < R$. *Moreover we have*

$$f^{(n)}(0) = n!\, a_n \quad (n = 0, 1, 2, \ldots). \qquad (5.31)$$

Proof This follows immediately upon repeated application of Theorem 3.

Example 3

It is perhaps surprising that the converse of the above corollary is false: a function f can be infinitely differentiable on an interval $|x| < R$ and yet not have a power series expansion valid on that interval. To see this, consider the function

$$f(x) = \begin{cases} e^{-1/x} & (x > 0), \\ 0 & (x \leq 0). \end{cases} \tag{5.32}$$

We showed in Section 3.5 that $f^{(k)}(0) = 0$ for all k. If we had the expansion $f(x) = \sum_0^\infty a_n x^n$, then we would have

$$a_n = \frac{1}{n!} f^{(n)}(0) = 0$$

for all n, and this would imply that $f(x) \equiv 0$. Therefore f has no power series expansion about $x = 0$.

A function that has a power series expansion about $x = x_0$, with positive radius of convergence, is said to be *analytic* at x_0. We conclude this section by showing that the exponential function e^x is analytic at every point; the same is true for the trigonometric functions $\sin x$ and $\cos x$; see Exercise 3.

The following theorem is a generalization of the mean value theorem. The proof, which is slightly technical, is given in the next section.

Theorem 4 (Taylor's theorem with remainder)   *Let f be n times continuously differentiable on the interval $|x - x_0| < h$. Then for all x in this interval we have*

$$f(x) = \sum_{k=0}^{n-1} \frac{f^{(k)}(x_0)}{k!} (x - x_0)^k + R_n(x), \tag{5.33}$$

where the "remainder" $R_n(x)$ is given by

$$R_n(x) = \frac{f^{(n)}(\xi)}{n!} (x - x_0)^n,$$

for some point ξ lying between x and x_0.

Corollary   *Let f be infinitely differentiable on the interval $|x - x_0| < h$. Then f has a power series expansion valid on this interval if and only if $R_n(x) \to 0$ as $n \to \infty$, for $|x - x_0| < h$, where $R_n(x)$ is given by (5.33).*

Proof If $R_n(x) \to 0$ as $n \to \infty$, then by letting $n \to \infty$ in (5.33) we obtain

$$f(x) = \sum_0^\infty \frac{f^{(k)}(x_0)}{k!}(x - x_0)^k. \tag{5.34}$$

Conversely if f has a power series expansion about x_0, it must have the form (5.34). Hence letting $n \to \infty$ in (5.33) we see that $\lim_{n\to\infty} R_n(x) = 0$. ∎

The series (5.34) is called the *Taylor series* of the function f, about the point x_0.

Example 4

$$e^x = \sum_0^\infty \frac{x^n}{n!} \quad \text{for all } x. \tag{5.35}$$

To see this, substitute $f(x) = e^x$ and $x_0 = 0$ in (5.33). Since $f^{(k)}(0) = 1$ for all k, we have

$$e^x = \sum_{k=0}^{n-1} \frac{x^k}{k!} + R_n(x),$$

where (for $x \neq 0$)

$$R_n(x) = \frac{e^\xi}{n!} x^n \quad (0 < |\xi| < |x|).$$

For each given x, therefore,

$$|R_n(x)| \leq e^{|x|} \frac{|x|^n}{n!} \to 0 \quad \text{as} \quad n \to \infty.$$

Consequently e^x has the power series expansion (5.35).

We can easily find the expansion of e^x about an arbitrary point x_0 by the following device:

$$e^x = e^{x_0}\, e^{x-x_0} = \sum_0^\infty \frac{e^{x_0}}{n!}(x - x_0)^n. \tag{5.36}$$

Formulas (5.35) and (5.36) are quite convenient for numerical computation of e^x; see Exercises 7 and 8.

Exercises

1. Find the Taylor series expansions about $x = 0$ for the following functions. Prove convergence.

   (a) $\sin x$;   (b) $\cos x$;   (c) $\sinh x = \frac{1}{2}(e^x - e^{-x})$.

2. If $i = \sqrt{-1}$, "prove" Euler's formula

$$e^{ix} = \cos x + i \sin x$$

   by finding power series expansions for both sides. Is this a valid proof?

3. Find the Taylor series expansion of $\sin x$ about an arbitrary point x_0.

4. Use the geometric series to derive explicit formulas for

   (a) $\displaystyle\sum_{n=1}^{\infty} n x^n$;   (b) $\displaystyle\sum_{n=1}^{\infty} n^2 x^n$.

5. Show that the series $\sum_1^{\infty} x^n/n^2$ converges uniformly on the interval $[-1, 1]$.

6. Generalize the result of Problem 5 to arbitrary series $\sum_0^{\infty} a_n x^n$ having radius of convergence $R = 1$.

7. Let x be given. Define two sequences $\{\delta_n\}$ and $\{y_n\}$ $(n = 0, 1, 2, \ldots)$ recursively as follows:

$$\delta_0 = x; \qquad \delta_n = \frac{x}{n+1} \delta_{n-1},$$

$$y_0 = 1; \qquad y_n = y_{n-1} + \delta_{n-1}.$$

   (a) Show that $y_n = \sum_{k=0}^{n} \dfrac{x^k}{k!}$.

   (b) Compute $e^{-0.1}$ to 5 decimals.
   (Cf. Table I of Chapter 1, page 37. Note also that the power series for e^x is alternating if $x < 0$.)

8. Use power-series methods to calculate $\int_0^1 e^{-t^2}\, dt$ to 3 decimals.

5.8   *Taylor's Theorem*

We will first prove the following version of Taylor's theorem, in which the remainder is in *integral form*.

Theorem 1   Let f be n times continuously differentiable on the interval $|x - x_0| < h$. Then for all x in this interval we have

$$f(x) = \sum_{k=0}^{n-1} \frac{f^{(k)}(x_0)}{k!} (x - x_0)^k + R_n(x),$$ (5.37)

where

$$R_n(x) = \frac{1}{(n-1)!} \int_{x_0}^{x} f^{(n)}(t)(x - t)^{n-1}\, dt.$$ (5.38)

Proof Let us use induction. For $n = 1$ the desired result is that if f is continuously differentiable then

$$f(x) = f(x_0) + \int_{x_0}^{x} f'(t)\, dt,$$

a result that is very familiar.

Assume therefore that (5.37) and (5.38) are valid for n. Integrating the remainder $R_n(x)$ by parts, we obtain

$$R_n(x) = \frac{1}{(n-1)!} \int_{x_0}^{x} f^{(n)}(t)(x - t)^{n-1}\, dt$$

$$= \frac{1}{(n-1)!} \left\{ \left[-f^{(n)}(t) \frac{(x - t)^n}{n} \right]_{x_0}^{x} + \frac{1}{n} \int_{x_0}^{x} f^{(n+1)}(t)(x - t)^n\, dt \right\}$$

$$= \frac{1}{n!} f^{(n)}(x_0)(x - x_0)^n + R_{n+1}(x).$$

Substitution of this into (5.37) proves the theorem for the case $n + 1$. This completes the induction. ∎

In order to obtain the usual form of the remainder in Taylor's formula, we require the following result.

Theorem 2 (Mean value theorem for integrals)   Let f, g be continuous functions on $[a, b]$ and suppose $g(x) \geq 0$ on $[a, b]$. Then there exists a point c in $[a, b]$ such that

$$\int_{a}^{b} f(x)g(x)\, dx = f(c) \int_{a}^{b} g(x)\, dx.$$ (5.39)

Proof Let $M = \max_{[a,b]} f(x)$ and $m = \min_{[a,b]} f(x)$. Since $g(x) \geq 0$ on $[a, b]$, we have

$$mg(x) \leq f(x)g(x) \leq Mg(x) \quad (x \in [a, b]),$$

and therefore

$$m \int_{a}^{b} g(x)\, dx \leq \int_{a}^{b} f(x)g(x)\, dx \leq M \int_{a}^{b} g(x)\, dx.$$

If $g(x) \equiv 0$ in $[a, b]$, the theorem is trivial. If $g(x) \not\equiv 0$ in $[a, b]$, then $\int_a^b g(x)\, dx > 0$, so that we can write

$$m \le \frac{\displaystyle\int_a^b f(x)g(x)\, dx}{\displaystyle\int_a^b g(x)\, dx} \le M.$$

Now since f is a continuous function on $[a, b]$, $f(x)$ must assume every value between m and M somewhere on $[a, b]$. Therefore there is a point c in $[a, b]$ such that

$$f(c) = \frac{\displaystyle\int_a^b f(x)g(x)\,dx}{\displaystyle\int_a^b g(x)\,dx}$$

and this is the same as (5.39). ∎

Notice that Theorem 2 is obviously also valid in the case that $g(x) \le 0$ throughout $[a, b]$. We can now derive our desired formula.

Theorem 3   Let f be n *times continuously differentiable on* $|x - x_0| < h$. *Then* (5.37) *is valid, with*

$$R_n(x) = \frac{f^{(n)}(c)}{n!}(x - x_0)^n \tag{5.40}$$

for some point c between x_0 and x.

Proof Suppose for simplicity that $x > x_0$. Then $g(t) = (x - t)^{n-1}$ is non-negative for $t \in [x_0, x]$. By Theorem 2 there exists a point c in $[x_0, x]$ such that

$$\int_{x_0}^x f^{(n)}(t)(x - t)^{n-1}\, dt = f^{(n)}(c) \int_{x_0}^x (x - t)^{n-1}\, dt$$

$$= f^{(n)}(c)\, \frac{(x - x_0)^n}{n}.$$

Hence (5.38) reduces to (5.40). The case $x < x_0$ is similar. ∎

Exercises

1. Prove the binomial theorem as a consequence of Taylor's theorem.

2. Use Taylor's theorem with integral form of the remainder to evaluate the integral $\int_0^x t^n e^t \, dt$.

3. Derive "Cauchy's" form of the remainder:

$$R_n(x) = \frac{f^{(n)}(c)}{(n-1)!}(x-c)^{n-1}(x-x_0)$$

(for some c between x_0 and x) by applying the mean value theorem for integrals with $g(t) \equiv 1$.

4. Show that the mean value theorem for derivatives is a consequence of the mean value theorem for integrals.

Note: The following exercises are concerned with the interpretation of an integral as an "average," or "mean."

5. If f is integrable over $[a, b]$, the number

$$\mu(f) = \frac{1}{b-a}\int_a^b f(x) \, dx$$

is called the *mean*, or *average*, of f over $[a, b]$. Show that $\mu(f)$ has the following properties:

  (i) $\mu(c) = c$ if $c = $ const.
  (ii) μ is linear: $\mu(\alpha f + \beta g) = \alpha\mu(f) + \beta\mu(g)$ (α, β constants).
  (iii) $f \le g$ implies $\mu(f) \le \mu(g)$.

6. Let $T(t)$ denote the temperature in Nome, Alaska, measured at a time t hours after midnight of December 31, 1946. How would you define, and how calculate, the "mean temperature" in Nome, Alaska during January 1947?

7. Explain why the name "mean value theorem" is appropriate for Theorem 2 in the case $g(x) \equiv 1$.

8. If $g(x) \ge 0$ on $[a, b]$, the number

$$\mu_g(f) = \frac{\displaystyle\int_a^b f(x)g(x)\,dx}{\displaystyle\int_a^b g(x)\,dx}$$

(assuming it exists) is called the *weighted mean of f* over [a, b], with *weight function g*. Show that the weighted mean of f satisfies the conditions (i)–(iii) of Exercise 5. Explain how Theorem 2 is a "mean value theorem" in the general case. (Assume f, g continuous.)

9. Let $g(x) = x$ and $[a, b] = [0, 1]$. Determine $c \in [0, 1]$ such that

$$\mu_g(f) = f(c) \quad \text{if } f(x) = e^x.$$

5.9   *On the Definition of the Exponential Function*

In previous sections we have assumed, for the purpose of examples, that you are familiar with the exponential function and its properties. From a strictly logical point of view this assumption is indefensible, inasmuch as a certain amount of "analysis" is required merely to define e^x. Fortunately we have now come far enough to give a rigorous definition and theory of e^x—and also a^x, log x, and so forth. In fact, we could define e^x in at least three completely different ways!

The most naive way to define e^x is first to define e^r for rational numbers r (by using the method of recursion formulas of Chapter 1, for example), and then to define e^x for irrational x as $\lim_{n \to \infty} e^{r_n}$, where $\{r_n\}$ is a sequence of rational numbers approaching x. We leave to your imagination how unpleasant it would be to use this approach to develop the theory of the exponential function.

A second method of defining e^x is to define log x first by means of an integral; then e^x can be defined as the inverse function of log x. This method is quite simple and elegant, and is used in many modern calculus texts.†

The method we will use is due to Weierstrass, and is based on the elementary theory of power series. Thus we will *define* $e^x = \sum_{n=0}^{\infty} x^n/n!$. Historically this is backward, but logically it is acceptable and convenient.

Definition 1   *For any real number x we define*

$$e^x = \sum_{n=0}^{\infty} \frac{x^n}{n!}.$$ (5.41)

---

† For example, see T. M. Apostol, *Calculus*, Vol. I, Blaisdell Publishing Company (1967).

From our theory of power series we know that the series (5.41) converges for all x. Moreover, we have, by (5.30),

$$\frac{d}{dx} e^x = \sum_{n=0}^{\infty} \frac{n x^{n-1}}{n!} = \sum_{n=1}^{\infty} \frac{x^{n-1}}{(n-1)!} = \sum_{n=0}^{\infty} \frac{x^n}{n!}.$$

This proves the following result.

Theorem 1   *The function e^x is differentiable for all x, and we have*

$$\frac{d}{dx} e^x = e^x. \tag{5.42}$$

Notice, by the way, that if we put $x = 1$ in Formula (5.41), we get $e^1 = \sum_0^{\infty} 1/n!$, and this *is* the number e, as we proved in Section 1.9. This justifies the use of the symbol e in the definition of e^x. To justify the use of the exponential notation, we should verify that for example, $e^2 = e \cdot e$. We can do even better.

Theorem 2   *For arbitrary x, y we have*

$$e^{x+y} = e^x \cdot e^y. \tag{5.43}$$

Proof   Consider y as a constant. Define the function

$$f(x) = e^{x+y} \cdot e^{-x}.$$

From (5.42) and elementary rules of calculus, we have

$$f'(x) = e^{x+y} \frac{d}{dx} (e^{-x}) + e^{-x} \frac{d}{dx} (e^{x+y})$$

$$= e^{x+y}(-e^{-x}) + e^{-x} e^{x+y} \equiv 0.$$

Consequently $f(x) \equiv C = $ constant, with

$$C = f(0) = e^y$$

(since $e^0 = 1$ by (5.41)). Hence we have shown that, for any y,

$$e^{x+y} \cdot e^{-x} = e^y. \tag{5.44}$$

Putting $y = 0$, we get $e^x \cdot e^{-x} = 1$. Thus, multiplying (5.44) on both sides by e^x, we obtain finally

$$e^{x+y} = e^x e^y. \quad \blacksquare$$

Corollary   *The function e^x is strictly increasing, and*

$$\lim_{x \to -\infty} e^x = 0; \qquad \lim_{x \to +\infty} e^x = +\infty.$$

Proof   From the definition (5.41) we see that $e^x > 0$ if $x \geq 0$. From Theorem 2, $e^{-x} = 1/e^x > 0$ if $x > 0$; thus $e^x > 0$ for all x. Since

$$\frac{d}{dx} e^x = e^x,$$

we conclude that e^x is strictly increasing. Now we know that $e > 2$, and from (5.43) it follows by induction that $e^n > 2^n \to +\infty$ as $n \to \infty$. The fact that e^x is increasing shows that $e^x \to +\infty$ as $x \to \infty$. Finally

$$\lim_{x \to -\infty} e^x = \lim_{x \to +\infty} e^{-x} = \lim_{x \to +\infty} \frac{1}{e^x} = 0. \quad \blacksquare$$

From the corollary we see that the function e^x has a uniquely determined inverse function, defined for $x > 0$:

Definition 2   *The function $\log x$, $x > 0$, is defined to be the inverse of the exponential function:*

$$\log x = t \quad \textit{if and only if} \quad x = e^t. \tag{5.45}$$

Theorem 3   *We have*

(a) $\dfrac{d}{dx} \log x = \dfrac{1}{x}$,

(b) $\log(xy) = \log x + \log y$.

Proof   (a) If $\log x = t$, we have by calculus (Equation 3.18)

$$\frac{dt}{dx} = \frac{1}{dx/dt} = \frac{1}{e^t} = \frac{1}{x}.$$

(b) Let $\log x = w$ and $\log y = z$. Then, by (5.45),

$$x = e^w \quad \text{and} \quad y = e^z.$$

Therefore,

$$xy = e^w e^z = e^{w+z},$$

which means that $\log (xy) = w + z = \log x + \log y$. ∎

Definition 3   *Let $a > 0$ be given. Define*

$$a^x = e^{x \log a}. \qquad (5.46)$$

The usual properties of a^x can immediately be deduced from (5.46); see Exercise 1.

Exercises

1. Prove the following formulas. $(x, y, \ldots$ denote arbitrary real numbers, whereas $a, b, \ldots$ denote positive real numbers.)

    (a) $(ab)^x = a^x b^x$,

    (b) $\log a^x = x \log a$,

    (c) $(a^x)^y = a^{xy}$,

    (d) $\dfrac{d}{dx} a^x = a^x \log a$,

    (e) $\dfrac{d}{dx} x^a = ax^{a-1}$.

2. Show that $\log x = \displaystyle\int_1^x \frac{dt}{t}$.

3. If $a > 0$, show that

    $$\log x < \frac{x^a}{a} \quad (x \geq 1).$$

    (*Hint*: Use Exercise 2.)

*4. If $a > 0, b > 0$, show that

    $$(a + b)^x \leq a^x + b^x \quad (0 \leq x \leq 1).$$

The following exercises indicate how the properties of the trigonometric functions can be derived from the power series for these functions.† Thus we define the functions $s(x)$ and $c(x)$ as follows.

$$s(x) = x - \frac{x^3}{3!} + \frac{x^5}{5!} - \cdots + (-1)^n \frac{x^{2n+1}}{(2n+1)!} + \cdots,$$

$$c(x) = 1 - \frac{x^2}{2!} + \frac{x^4}{4!} - \cdots + (-1)^n \frac{x^{2n}}{(2n)!} + \cdots.$$

5. Verify that $s'(x) = c(x)$ and $c'(x) = -s(x)$.

6. Show that
$$s(x + a) = s(x)c(a) + c(x)s(a),$$

$$c(x + a) = c(x)c(a) - s(x)s(a).$$

(*Hints*: (a) Let $f(x) = s(x + a) - s(x)c(a) - c(x)s(a)$. Show that $f'' + f = 0$. (b) Hence show that $[(f')^2 + f^2]' = 0$, so that $(f')^2 + f^2 = $ const. $= 0$. (c) Deduce the above formulas.)

7. Show that $(s(x))^2 + (c(x))^2 \equiv 1$.

The following exercises pertain to the *binomial series*.

8. Find Taylor's series (about $x = 0$) for the function
$$f(x) = (1 + x)^a$$

where $a \in \mathbb{R}$ is given.

9. Show that the Taylor series for $(1 + x)^a$ converges for $|x| < 1$.

10. Use the Lagrange form of the remainder $R_n(x)$, Equation (5.40), to show that the Taylor series for $(1 + x)^a$ converges to $(1 + x)^a$ for $0 \le x < 1$. (*Note:* The Cauchy form of the remainder, Exercise 3, Section 5.8, can be used to obtain the same result for $-1 < x < 0$, but the proof is slightly more complicated.)‡

---

† For a complete treatment, see K. Knopp, *Theory and Application of Infinite Series*, Hafner (1951).

‡ See A. E. Taylor and W. R. Mann, *Advanced Calculus*, 2nd ed., Xerox (1972), p. 630.

6 *Limits and Continuity in n Dimensions*

In this chapter we wish to generalize the theory of limits and continuity to deal with functions of several variables. For the most part, the generalization is very natural and simple. It depends primarily on the observation that limits are defined in terms of distances between points: for example, the inequality $|x - a| < \delta$ is a statement about the distance between x and a. The definition of distance generalizes easily to n dimensions.

In discussing functions of one variable, we invariably considered functions defined on an interval I. Although the notion of an interval generalizes easily to n dimensions (for example, a rectangular "box"), it is obvious that if our theory is going to be useful we must consider a much wider class of n-dimensional objects than just "intervals." Not only do we wish to consider arbitrarily shaped "solid" regions, but we are also interested in more complicated objects, such as "surfaces," or "curves." Such objects have no interesting counterpart when $n = 1$.

A major complication, then, in passing from one dimension to n dimensions, lies in the nature of the subsets of n-dimensional space that we consider. Fortunately this leads to only relatively minor complications in the study of limits and continuity. It leads to tremendous additional complications, however, in the study of the calculus itself. These complications, treated in texts on advanced calculus, vector analysis, differential geometry, and so on, are beyond the scope of this text.

6.1   *Open and Closed Sets in* $\mathbb{R}^n$

A *point* in Euclidean *n*-space is, by definition, an ordered *n*-tuple of real numbers:†

$$x = (x^1, x^2, \ldots, x^n). \tag{6.1}$$

The number x^j is called the *j*th *coordinate* of *x*. The set of all such points is denoted by $\mathbb{R}^n$, and $\mathbb{R}^n$ is called *Euclidean n-space*.

It is convenient also to introduce a "linear structure" on $\mathbb{R}^n$ by defining *addition* and *scalar multiplication* as follows:

$$x + y = (x^1 + y^1, x^2 + y^2, \ldots, x^n + y^n) \quad (x, y \in \mathbb{R}^n) \tag{6.2}$$

$$\alpha x = (\alpha x^1, \alpha x^2, \ldots, \alpha x^n) \quad (x \in \mathbb{R}^n, \alpha \in \mathbb{R}). \tag{6.3}$$

Geometrically speaking, it is useful to visualize the points *x* in $\mathbb{R}^n$ either as points, or as *vectors* pointing from the origin $0 = (0, 0, \ldots, 0)$ to *x* (Figure 6.1). If we use the vector interpretation, then Equations (6.2) and (6.3), respectively, become the usual parallelogram law and scalar multiplication law for vectors (see Figure 6.2).

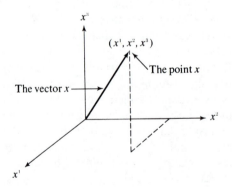

Figure 6.1   Points vs. vectors in $\mathbb{R}^n (n = 3)$.

---

† We use the superscript notation for coordinates and reserve subscripts for use with sequences.

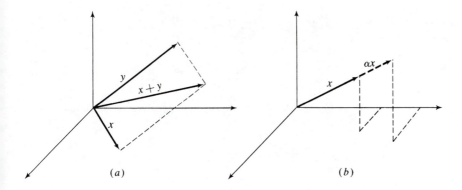

(a) Parallelogram law of addition   (b) Scalar multiplication

Figure 6.2   Vector operations ($n = 3$).

The following rules of calculation are immediately obvious from the definitions of Formulas (6.2) and (6.3).

Properties of vector operations in $\mathbb{R}^n$

1. $x + y = y + x.$

2. $x + (y + z)$
    $\qquad = (x + y) + z.$

3. $\alpha(\beta x) = (\alpha\beta)x.$

4. $\alpha(x + y) = \alpha x + \alpha y.$

5. $(\alpha + \beta)x = \alpha x + \beta x.$

6. $x + 0 = x,$
    where $0 = (0, 0, \ldots, 0).$

7. $x + (-x) = 0,$
    where $-x = (-x^1, -x^2, \ldots, -x^n).$

8. $0x = 0$ and $1x = 1.$

Definition 1   *If $x \in \mathbb{R}^n$, we define the **length** of x as*

$$\|x\| = \left\{ \sum_{i=1}^{n} (x^i)^2 \right\}^{1/2}. \tag{6.4}$$

*If $x, y \in \mathbb{R}^n$, we define **the distance between** x and y as the length of $x - y$:*

$$\|x - y\| = \left\{ \sum_{i=1}^{n} (x^i - y^i)^2 \right\}^{1/2}. \tag{6.5}$$

It is of basic importance that this distance satisfies the triangle inequality. To prove this we need the following lemma.

Lemma 1 (Schwarz's inequality)   *For any* $x, y \in \mathbb{R}^n$,

$$\left| \sum_{i=1}^{n} x^i y^i \right| \leq \|x\| \, \|y\|. \tag{6.6}$$

Proof Let x, y be given, and let α denote an arbitrary real parameter. Then

$$0 \leq \|x + \alpha y\|^2 = \sum_{i=1}^{n} (x^i + \alpha y^i)^2$$

$$= \sum_{i=1}^{n} (x^i)^2 + 2\alpha \sum_{i=1}^{n} x^i y^i + \alpha^2 \sum_{i=1}^{n} (y^i)^2$$

$$= A + 2B\alpha + C\alpha^2.$$

Now by elementary algebra we know that if the quadratic expression $A + 2B\alpha + C\alpha^2$ is ≥ 0 for all values of α, then $B^2 - AC \leq 0$. Therefore,

$$\left(\sum x^i y^i \right)^2 - \sum (x^i)^2 \cdot \sum (y^i)^2 \leq 0.$$

This is exactly equivalent to (6.6). ∎

Definition 2   *If x and y are vectors in $\mathbb{R}^n$, the expression*

$$x \cdot y = \sum_{i=1}^{n} x^i y^i$$

*is called the **dot product** (or **inner product**) of x and y.*

Schwarz's inequality can therefore be written as

$$|x \cdot y| \leq \|x\| \, \|y\|.$$

Angles can also be introduced into $\mathbb{R}^n$ by means of the formula

$$\cos \theta = \frac{x \cdot y}{\|x\| \, \|y\|},$$

which determines a unique angle θ, $0 \leq \theta \leq \pi$, called the *angle between* the vectors x and y.

Theorem (Triangle inequality)   *For any* $x, y \in \mathbb{R}^n$

$$\|x + y\| \le \|x\| + \|y\|. \tag{6.7}$$

Proof Consider

$$\begin{aligned}
\|x + y\|^2 &= \sum (x^i + y^i)^2 \\
&= \sum (x^i)^2 + 2 \sum x^i y^i + \sum (y^i)^2 \\
&\le \|x\|^2 + 2\|x\|\,\|y\| + \|y\|^2 \quad \text{(by Schwarz's inequality)} \\
&= (\|x\| + \|y\|)^2. \quad \blacksquare
\end{aligned}$$

Corollary   *For any* 3 *points* x, y, z *in* $\mathbb{R}^n$,

$$\|x - y\| \le \|x - z\| + \|z - y\|. \tag{6.8}$$

The proof is obvious. Geometrically the inequality (6.8) is indeed a fact about triangles, see Figure 6.3.

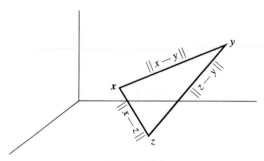

Figure 6.3

Definition 3   *Let* x_0 *be a given point of* $\mathbb{R}^n$ *and let* $r > 0$. *Then the set* $N_r(x_0)$, defined by

$$N_r(x_0) = \{x \in \mathbb{R}^n | \ \|x - x_0\| < r\}, \tag{6.9}$$

is called the **(open) ball of radius** r **centered at** x_0. *The set* $S_r(x_0)$, *defined by*

$$S_r(x_0) = \{x \in \mathbb{R}^n | \ \|x - x_0\| = r\} \tag{6.10}$$

is called the **sphere of radius** r **centered at** x_0.

The ball $N_r(x_0)$ is also sometimes called the *r*-**neighborhood** of x_0. Also, the set

$$N_r^*(x_0) = \{x \in \mathbb{R}^n | 0 < \|x - x_0\| < r\}$$

is called the **deleted** *r*-**neighborhood** of x_0.

Definition 4   *Let D be a given subset of* $\mathbb{R}^n$. *We say that D is an* **open** *set provided that every point x of D is the center of some ball* $N_r(x)$, $r > 0$, *contained within D.*

The whole space $D = \mathbb{R}^n$ is obviously an open set; so is the empty set, $\varnothing$ (since $\varnothing$ has *no* points, the condition of Definition 4 is surely satisfied!). Another easy example of an open set is the *half-space*

$$H = \{x \in \mathbb{R}^n \mid x^1 > 0\}.$$

Lemma 2   *The ball* $N_r(x_0)$ *is always an open set.*

Proof   Let $z_0 \in N_r(x_0)$. Then $\|x_0 - z_0\| = p < r$. We can show (Figure 6.4) that $N_{r-p}(z_0)$ is contained in $N_r(x_0)$. For if $x \in N_{r-p}(z_0)$, then $\|x - z_0\| < r - p$, so

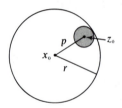

Figure 6.4

that, by the triangle inequality,

$$\|x - x_0\| \le \|x - z_0\| + \|z_0 - x_0\| < r - p + p = r,$$

that is, $x \in N_r(x_0)$. This proves that $N_r(x_0)$ is open.   ∎

Notice that when $n = 1$, the ball $N_r(x_0)$ is the same as the open interval $(x_0 - r, x_0 + r)$.

Definition 5   *A subset D of* $\mathbb{R}^n$ *is called* **closed** *if the "complementary set"* $D' = \{x \in \mathbb{R}^n \mid x \notin D\}$ *is an open set.*

Obvious examples of closed sets are $\mathbb{R}^n$ itself (whose complement is the open set $\varnothing$), the empty set, $\varnothing$ (whose complement is $\mathbb{R}^n$), and the closed half-space

$$\bar{H} = \{x \in \mathbb{R}^n \mid x^1 \ge 0\}.$$

Another example is the *closed ball*

$$\bar{N}_r(x_0) = \{x \in \mathbb{R}^n |\ \|x - x_0\| \leq r\}$$

(see Exercise 3). In particular, a closed interval $[a, b]$ in $\mathbb{R}$ is a closed set.

Let us point out that a given subset of $\mathbb{R}^n$ need not be either open or closed. For example, a half-open interval $(a, b]$ in $\mathbb{R}$ fits neither definition.

One further notion which is useful in studying continuity is that of the boundary of a set.

Definition 6   *Let D be a subset of* $\mathbb{R}^n$ *and let* $x_0 \in \mathbb{R}^n$. *Then* x_0 *is called a* **boundary point** *of D provided every ball* $N_r(x_0)$ *contains some point of D and some point of the complement, D′ (see Figure 6.5). By the* **boundary** *of D we mean the set of boundary points of D.*

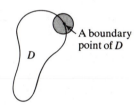

Figure 6.5

Example

The boundary of the ball $N_r(z_0)$ is the sphere $S_r(z_0)$. This is intuitively obvious and is easily verified by considering the straight line L through z_0 and $x_0 \in S_r(z_0)$. Any ball $N_p(x_0)$ contains points on L both inside and outside $N_r(z_0)$ (Figure 6.6).

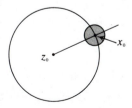

Figure 6.6

Exercises

1. Let $x = (3, 1, 0)$ and $y = (1, -1, 2)$ be points in $\mathbb{R}^3$. Calculate $\|x\|$, $\|y\|$, $\|x + y\|$, and $\sum x^i y^i$, and check that the Schwarz and triangle inequalities are valid.

2. Prove the inequality

    $$\left(\sum_{i=1}^{n} x^i \right)^2 \leq n \sum_{i=1}^{n} (x^i)^2.$$

3. Show that $\bar{N}_r(x_0) = \{x \in \mathbb{R}^n | \ \|x - x_0\| \leq r\}$ is a closed set.

4. Decide whether the following sets are open or closed or neither.

    (a) $\{x \in \mathbb{R}^3 \,|\, x^1 = 0\}$;

    (b) A set consisting of a single point;

    (c) $\{x \in \mathbb{R}^n \,|\, x \neq 0\}$;

    (d) $\{x \in \mathbb{R} \,|\, x = 1/k$ for some integer $k = 1, 2, 3, \ldots\}$;

    (e) $\{x \in \mathbb{R}^2 \,|\, x^1$ is rational$\}$;

    (f) $\bar{N}_r(0) \cap \bar{H} = \{x \in \mathbb{R}^n | \|x\| \leq r \quad \text{and} \quad x^1 \geq 0\}$.

5. Prove that the intersection of two open sets D_1 and D_2 is an open set.

6. Prove that the intersection of a finite number (≥ 2) of open sets is an open set.

7. Show that the intersection of an infinite number of open sets need not be an open set.

8. Prove that the union of any number (finite or infinite) of open sets is an open set.

9. Prove in detail that $\{x \in \mathbb{R}^n \,|\, x^1 = 0\}$ is closed.

10. Show that a set D is closed if and only if it contains all its boundary points.

11. Show that the union of a set D with the set of its boundary points is a closed set. (This set is called the closure $\bar{D}$ of D.)

12. Show that

    (a) the union of finitely many closed sets is a closed set,

    (b) the intersection of an arbitrary family of closed sets is a closed set.

    (Use the results of Exercises 6 and 8; see also Exercise 3 of Appendix I.4.)

6.2 Sequences and Limits

The limit of a sequence $\{x_j\}$ of points in $\mathbb{R}^n$ is defined by an obvious adjustment to the definition of Chapter 1.

Definition   *Let* $\{x_j\}$ *be a sequence of points in* $\mathbb{R}^n$, *and let* $x_0 \in \mathbb{R}^n$. *We say that* $\{x_j\}$ *converges to* x_0; *symbolically,*

$$\lim_{j \to \infty} x_j = x_0 \quad [\text{or } x_j \to x_0 \text{ as } j \to \infty],$$

provided the following condition is satisfied:

given $\varepsilon > 0$, *there exists an integer* N *such that*

$$\|x_j - x_0\| < \varepsilon \quad \text{for all } j \geq N. \tag{6.11}$$

Notice that $\lim_{j \to \infty} x_j = x_0$ if and only if $\lim_{j \to \infty} \|x_j - x_0\| = 0$.

Lemma 1   *The sequence* $\{x_j\}$ *converges to* x_0 *if and only if each of the coordinate sequences* $\{x_j^k\}_{j=1}^\infty$ *converges to* x_0^k, *for* $k = 1, 2, \ldots, n$.

Proof   First, if $x_j \to x_0$, then for each $k = 1, 2, \ldots, n$,

$$|x_j^k - x_0^k| \leq \left\{ \sum_{i=1}^n |x_j^i - x_0^i|^2 \right\}^{1/2}$$

$$= \|x_j - x_0\| \to 0 \quad \text{as} \quad j \to \infty,$$

so that $x_j^k \to x_0^k$. Conversely, if $x_j^k \to x_0^k$ for each $k = 1, 2, \ldots, n$, then

$$\|x_j - x_0\| = \left\{ \sum_{k=1}^n |x_j^k - x_0^k|^2 \right\}^{1/2} \to 0 \quad \text{as} \quad j \to \infty,$$

so that $x_j \to x_0$. ∎

The reason for the adjective "closed" in the term "closed set" is explained by the following simple result.

Lemma 2   *Let D be a subset of $\mathbb{R}^n$. Then D is closed if and only if every convergent sequence of points in D converges to a point of D.*

Proof   Suppose D is closed. Let $\{x_j\}$ be a sequence of points of D, with $\lim_{j \to \infty} x_j = x_0$. If $x_0 \notin D$, then $x_0 \in D'$. Since D' is open, there exists a ball $N_r(x_0)$, $r > 0$, contained in D'. But each x_j is in D, so that $\|x_j - x_0\| \geq r$, which contradicts the assumption that $x_j \to x_0$. Therefore $x_0 \in D$.

Conversely, assume that every convergent sequence of points in D converges to a point of D. Suppose that D is not closed; in other words, D' is not open. Then there is a point x_0 in D' such that no ball $N_r(x_0)$, $r > 0$, is contained in D'. Thus every ball $N_{1/j}(x_0)$, $j = 1, 2, 3, \ldots$, contains some point x_j of D. Since $\|x_j - x_0\| < 1/j$, the sequence $\{x_j\}$ converges to $x_0 \notin D$. This is a contradiction, and therefore D must be closed.   ∎

Example

We can easily check that the set

$$\bar{H} = \{x \in \mathbb{R}^n \mid x_1 \geq 0\}$$

is closed by using Lemma 2. For if $\{x_j\}$ is a sequence in $\bar{H}$ with $x_j \to x_0$, then $x_j^1 \geq 0$ for all j, so that (by Lemma 1) $x_0^1 = \lim_{j \to \infty} x_j^1 \geq 0$. This means that $x_0 \in \bar{H}$, and therefore $\bar{H}$ must be closed.

The following simple properties of limits of sequences in $\mathbb{R}^n$ can be easily proved.

Theorem 1

(a)   *If $\{x_j\}$ converges, the limit is unique.*

(b)   *If $\{x_j\}$ converges and α is a constant, then*

$$\lim_{j \to \infty} \alpha x_j = \alpha \lim_{j \to \infty} x_j.$$

(c)   *If $\{x_j\}$ and $\{y_j\}$ converge, then*

$$\lim_{j \to \infty} (x_j + y_j) = \lim_{j \to \infty} x_j + \lim_{j \to \infty} y_j.$$

Next we come to the Cauchy criterion for convergence. A sequence $\{x_j\}$ of points of $\mathbb{R}^n$ is called a *Cauchy sequence* provided that:

given $\varepsilon > 0$, there exists an integer N such that

$$\|x_j - x_k\| < \varepsilon \quad \text{for all } j, k \geq N. \tag{6.12}$$

The condition (6.12) can also be expressed by writing

$$\lim_{j,k \to \infty} \|x_j - x_k\| = 0. \tag{6.13}$$

Theorem 2 (Cauchy's criterion)   *A given sequence $\{x_j\}$ in $\mathbb{R}^n$ converges if and only if it is a Cauchy sequence.*

Proof Suppose $\{x_j\}$ is a Cauchy sequence. Then, as in Lemma 1, each coordinate sequence $\{x_j^l\}$, $l = 1, 2, \ldots, n$ is also a Cauchy sequence, since $|x_j^l - x_k^l| \leq \|x_j - x_k\|$. Hence, $\lim_{j \to \infty} x_j^l = x_0^l$ exists for each l. By Lemma 1,

$$\lim_{j \to \infty} x_j = x_0.$$

Conversely, if $\lim_{j \to \infty} x_j = x_0$, then each coordinate sequence converges, and is therefore a Cauchy sequence:

$$\lim_{j,k \to \infty} |x_j^l - x_k^l| = 0 \quad (l = 1, 2, \ldots, n).$$

Consequently also

$$\|x_j - x_k\| = \left\{ \sum_{l=1}^n |x_j^l - x_k^l|^2 \right\}^{1/2} \to 0 \quad \text{as} \quad j, k \to \infty;$$

that is, $\{x_j\}$ is a Cauchy sequence. ∎

We prove next the n-dimensional version of the Bolzano-Weierstrass theorem. If D is a given subset of $\mathbb{R}^n$, we call a point x_0 in $\mathbb{R}^n$ an *accumulation point* of D if every deleted neighborhood $N_r^*(x_0)$, $r > 0$, contains some point of D. It is clear that x_0 is an accumulation point of D if and only if there is a sequence $\{x_j\}$ of points of D, with $x_j \neq x_0$, which converges to x_0.

A point of D that is not an accumulation point of D is called an *isolated point* of D.

Theorem 3 (Bolzano-Weierstrass)   *Every infinite, bounded subset of $\mathbb{R}^n$ has an accumulation point in $\mathbb{R}^n$.*

Proof If D is bounded, it is contained in some cube Q_0 of side r_0. Divide Q_0 into 2^n congruent cubes of side $(1/2)r_0$, by bisecting each edge of Q_0. Let Q_1 denote one such cube containing infinitely many points of D.

Starting with Q_1, we obtain by the same process a sub-cube Q_2 of side $\frac{1}{4}r_0$, containing infinitely many points of D, and so on. Let $\{x_j\}$ be a sequence of points of D with $x_j \in Q_j$ for all j. Then $\{x_j\}$ is a Cauchy sequence, because if $j, k \geq N$, then x_j and x_k belong to Q_N, so that

$$\|x_j - x_k\| \leq \text{diam } Q_N = \frac{1}{2^N} \text{diam } Q_0 < \varepsilon \quad \text{(given)}$$

for sufficiently large N. By Theorem 2 the sequence $\{x_j\}$ converges to some point x_0 in $\mathbb{R}^n$. ∎

Corollary   *Every bounded sequence in $\mathbb{R}^n$ has a convergent subsequence.*

Exercises

1. Let $x_j = \left(\dfrac{j}{j+1}, \dfrac{1}{3^j} \right) \in \mathbb{R}^2 \quad (j = 1, 2, 3, \ldots)$.

    (a) Find $\lim_{j \to \infty} x_j = x_0$.

    (b) Given $\varepsilon > 0$, determine an integer N such that

    $$\|x_j - x_0\| < \varepsilon \quad \text{for all } j \geq N.$$

2. Prove Theorem 1(c).

3. Given $\lim_{j \to \infty} x_j = x_0$ and $\lim_{j \to \infty} y_j = y_0$, prove that

    $$\lim_{j \to \infty} x_j \cdot y_j = x_0 \cdot y_0.$$

4. If $x_j = (\sin(j\pi/2), \cos(j\pi/3))$, show that $\{x_j\}$ is bounded and find a convergent subsequence.

5. Find the set of accumulation points of the sets of Exercise 4, Section 6.1.

6. Express the statement "D is a closed set" in terms of the accumulation points of D, and prove your statement.

7. Show that if $\lim_{j \to \infty} x_j = x_0$, then

    $$\|x_j - z_0\| \to \|x_0 - z_0\| \quad \text{as } j \to \infty$$

for any fixed point z_0. (*Hint:* Use the "backward" triangle inequality $\|a - b\| \geq |\, \|a\| - \|b\| \,|$ in $\mathbb{R}^n$.)

8. (a) If $\{x_j\}$ is a given sequence in $\mathbb{R}^n$, define $\sum_{j=1}^{\infty} x_j$.

   (b) Prove that $\sum_{j=1}^{\infty} x_j$ converges if $\sum_{j=1}^{\infty} \|x_j\|$ does so. (Use the Cauchy criterion!)

   (c) Prove or disprove the converse to (b).

9. A set S in $\mathbb{R}^n$ is said to be *dense* if every nonempty open set D in $\mathbb{R}^n$ contains a point of S.

   (a) Show that S is dense if and only if every point of $\mathbb{R}^n$ is an accumulation point of S.

   (b) Show that there is a countable, dense subset of $\mathbb{R}^n$.

6.3   Limits and Continuity

In advanced calculus it is customary to begin by studying real-valued functions $f(x^1, x^2, \ldots, x^n)$ of several variables. In this text we immediately go one step further and discuss *vector-valued* functions of several variables. This means that the values

$$f(x) = f(x^1, x^2, \ldots, x^n)$$

will themselves be vectors† (or points) in another space $\mathbb{R}^m$:

$$f(x) = (f^1(x), f^2(x), \ldots, f^m(x)).$$

Example 1

Consider the function $f = (f^1, f^2, f^3)$ determined by the equations

$$\left.\begin{aligned}
f^1(\theta, \phi) &= \cos\theta \sin\phi, \\
f^2(\theta, \phi) &= \sin\theta \sin\phi, \\
f^3(\theta, \phi) &= \cos\phi.
\end{aligned}\right\} \tag{6.14}$$

Then f is a function defined on $\mathbb{R}^2$ with values in $\mathbb{R}^3$. We symbolize this fact by writing $f: \mathbb{R}^2 \to \mathbb{R}^3$. (The geometric interpretation of this function will be discussed later on.)

---

† In many texts special notations, such as arrows, bars, or boldface type, are used for vectors. Such notations, although sometimes convenient, are not necessary if the context is clearly understood.

Definition 1   *Let D be a given subset of $\mathbb{R}^n$. We use the notation*

$$f : D \to \mathbb{R}^m$$

*to indicate that f is a function defined on D and having its values in $\mathbb{R}^m$. The set D (which henceforth will usually be specified) is called the **domain** of the function f. The set of values of f, namely $\{f(x) \mid x \in D\}$, is called the **image of D under f** and is denoted by $f[D]$.*

The excuse for introducing such general functions at the outset is, first, that there are many important applications, and second, that questions about limits and continuity are easily handled for the general case.

Definition 2   *Let D be a subset of $\mathbb{R}^n$ and let $f : D \to \mathbb{R}^m$. If x_0 is an accumulation point of D, and $a \in \mathbb{R}^m$, we say that*

$$\lim_{x \to x_0} f(x) = a$$

provided the following condition is satisfied:

for every $\varepsilon > 0$ there exists $\delta > 0$ such that

$$\|f(x) - a\| < \varepsilon \quad \text{whenever } x \in D \text{ and } 0 < \|x - x_0\| < \delta. \tag{6.15}$$

Notice the similarity of this definition with the definition of limit for an ordinary function of one variable (Section 3.2). There is one subtle difference, however. Namely, we are not assuming that f is defined in some whole *neighborhood $N_r(x_0)$*. Hence, we have added the stipulation "$x \in D$" to the condition (6.15). (This subtlety could have been introduced in Chapter 3, but was avoided in order to keep things simple.)

Recall the numerical practice you had in Chapter 3 of calculating δ for given ε. Here is a simple 2-dimensional example for you to practice on (don't be discouraged if it turns out to be tricker than it looks!).

Example 2

Let $f : \mathbb{R}^2 \to \mathbb{R}^1$ be given by

$$f(x, y) = (x + 2y)^2.$$

Obviously $\lim_{(x,y) \to (3,2)} f(x, y) = 49$. Given $\varepsilon > 0$, determine δ so that (6.15) holds.

The following theorem ought to be patently obvious, and we state it only roughly.

Theorem 1

(a) *Limits are unique if they exist.*

(b) $\lim\limits_{x\to x_0} (f(x) + g(x)) = \lim\limits_{x\to x_0} f(x) + \lim\limits_{x\to x_0} g(x)$, *if these exist.*

(c) $\lim\limits_{x\to x_0} cf(x) = c \lim\limits_{x\to x_0} f(x)$, *if this exists.*

(d) $\lim\limits_{x\to x_0} (f(x) \cdot g(x)) = (\lim\limits_{x\to x_0} f(x)) \cdot (\lim\limits_{x\to x_0} g(x))$, *if these exist.*

We can now define continuity in terms of limits.

Definition 3   *Let $f: D \to \mathbb{R}^m$ and let $x_0 \in D$ be an accumulation point of D. Then f is said to be **continuous at** x_0 if*

$$\lim_{x\to x_0} f(x) = f(x_0).$$

*(A function $f: D \to \mathbb{R}^m$ is automatically considered to be continuous at any isolated points of D.) Furthermore, f is said to be **continuous on** D if it is continuous at each point of D.*

In ε—δ terms we have: $f: D \to \mathbb{R}^m$ *is continuous at $x_0 \in D$ if and only if, for every $\varepsilon > 0$ there exists $\delta > 0$ such that*

$$\|f(x) - f(x_0)\| < \varepsilon \quad \text{whenever } x \in D \text{ and } \|x - x_0\| < \delta.$$

Note that this applies to both accumulation points and isolated points of D.

Theorem 2   *Let $f, g: D \to \mathbb{R}^m$, and let $x_0 \in D$. If f and g are continuous at x_0, then so are the functions $f + g$, αf ($\alpha = $ constant) and $f \cdot g$.*[†]

Theorem 3   *Let $f: D \to \mathbb{R}^m$ and let $g: E \to \mathbb{R}^k$, where we suppose that the image of f is contained in E. If f is continuous at $x_0 \in D$ and g is continuous at $y_0 = f(x_0)$, then the composite function $g \circ f: D \to \mathbb{R}^k$ is continuous at x_0.*

† Note that $f \cdot g: D \to \mathbb{R}^1$.

Proof (*cf. Section* 3.4) Let $\varepsilon > 0$. Then there exists $\delta_1 > 0$ such that

$$\|g(y) - g(y_0)\| < \varepsilon \quad \text{for all } y \in E \text{ with } \|y - y_0\| < \delta_1.$$

Also, there exists $\delta_2 > 0$ such that

$$\|f(x) - f(x_0)\| < \delta_1 \quad \text{for all } x \in D \text{ with } \|x - x_0\| < \delta_2.$$

Hence, $x \in D$ with $\|x - x_0\| < \delta_2$ implies that

$$\|g(f(x)) - g(f(x_0))\| < \varepsilon. \quad \blacksquare$$

The following result states that *addition and multiplication are continuous operations.*

Theorem 4   *The functions*

$$f_1(x, y) = x + y, \qquad f_2(x, y) = xy$$

are continuous as functions of two variables (that is, as functions defined in $\mathbb{R}^2$).

Proof   Let $\phi(x, y) = x$ and $\Psi(x, y) = y$. It is obvious from the definition that ϕ and Ψ are everywhere continuous. Since $f_1 = \phi + \Psi$ and $f_2 = \phi \cdot \Psi$, Theorem 2 implies that f_1 and f_2 are continuous. $\quad \blacksquare$

Example 3

Consider the function

$$f(x, y) = \frac{xy}{x^2 + y^2} \quad (x^2 + y^2 \neq 0).$$

Obviously this function is continuous everywhere except $(0, 0)$, where it is not defined. To study the behavior of f near the origin, we will show that

(i) $\lim\limits_{(x,y) \to (0,0)} f(x, y)$ does not exist, although

(ii) $f(x, y)$ is bounded; in fact $|f(x, y)| < \frac{1}{2}$ for all x, y.

The proof of (ii) is easy: use the elementary inequality $2\,|xy| \leq x^2 + y^2$. To check (i), consider points (x, ax), where $a = \text{const}$. We have

$$f(x, ax) = \frac{a}{1 + a^2},$$

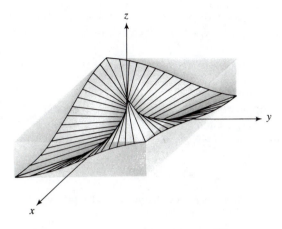

Figure 6.7   The surface $z = \dfrac{xy}{x^2 + y^2}$

which is constant (see Figure 6.7). Now $ax \to 0$ as $x \to 0$, so that if

$$\lim_{(x,y)\to(0,0)} f(x, y)$$

exists, it must equal

$$\lim_{x\to 0} f(x, ax) = \frac{a}{1 + a^2}$$

for any a. But this value varies with a; for example, it equals 0 if $a = 0$ and $1/2$ if $a = 1$. Hence (i) is proved.

The definition of continuity used so far has been in terms of limits. However there is another important characterization of continuity in terms of "inverse images" of open sets, which we give now. Given the function f, by the *inverse image* of a set U, we mean the set

$$f^{-1}[U] = \{x \in \text{domain of } f \mid f(x) \in U\}.$$

Theorem 5   *Let D be an open set in $\mathbb{R}^n$ and let $f : D \to \mathbb{R}^m$. Then f is continuous on D if and only if the inverse image $f^{-1}[U]$ of every open set U in $\mathbb{R}^m$ is an open set in $\mathbb{R}^n$.*

Proof   First, suppose f is continuous on D. Let U be an open set in $\mathbb{R}^m$ and let $x_0 \in f^{-1}[U]$. Then $y_0 = f(x_0) \in U$, and since U is open, there is a neighborhood $N_\varepsilon(y_0) \subset U$. Since f is continuous at x_0, there is a number $\delta > 0$ such that

$\|f(x) - f(x_0)\| < \varepsilon$, provided $\|x - x_0\| < \delta$ and $x \in D$. By taking δ smaller if necessary, we ensure that $N_\delta(x_0) \subset D$ (since D is open). Therefore

$$N_\delta(x_0) \subset f^{-1}[N_\varepsilon(y_0)] \subset f^{-1}[U],$$

and hence we have shown that $f^{-1}[U]$ is open.

Conversely, suppose $f^{-1}[U]$ is open for every open set $U \subset \mathbb{R}^m$. Let $x_0 \in D$ and let $y_0 = f(x_0)$. Then $f^{-1}[N_\varepsilon(y_0)]$ is open, for any $\varepsilon > 0$, so that there exists some neighborhood $N_\delta(x_0) \subset f^{-1}[N_\varepsilon(y_0)]$. This implies that

$$\|f(x) - y_0\| = \|f(x) - f(x_0)\| < \varepsilon$$

whenever $\|x - x_0\| < \delta$. Thus f is continuous at any point $x_0 \in D$. ∎

Example 4

Let $f(x, y) = x^2 + 4y^2$ be defined on $\mathbb{R}^2$. Then the inequality

$$x^2 + 4y^2 < 4$$

determines an open set in $\mathbb{R}^2$, since this set is just the inverse image of the open interval $(-\infty, 4)$.

It can also be easily shown that the inequality

$$x^2 + 4y^2 \leq 4$$

determines a *closed* set in $\mathbb{R}^2$. A similar situation prevails for any continuous function $f : \mathbb{R}^n \to \mathbb{R}$.

We conclude this section with some examples of functions from $\mathbb{R}^n$ to $\mathbb{R}^m$, $m \neq 1$.

Definition 4   *Let x_1, x_2 be points of $\mathbb{R}^m$ and let $I = [\alpha, \beta] \subset \mathbb{R}^1$. Any continuous function $f : I \to \mathbb{R}^m$ such that $f(\alpha) = x_1$ and $f(\beta) = x_2$ is called a **curve in** $\mathbb{R}^m$, **joining** x_1 **and** x_2.*†

---

† This is the modern definition of a curve as a continuous function. It may disagree slightly with the intuitive notion of a curve as a set of points, which now becomes the *image* of the curve.

Example 5

Consider the curve $f:[0, 2\pi] \rightarrow \mathbb{R}^3$ given by

$$f^1(t) = \cos t,$$

$$f^2(t) = \sin t,$$

$$f^3(t) = t.$$

This is a *helix* lying on the cylinder $(x^1)^2 + (x^2)^2 = 1$ and joining the points $(1, 0, 0)$ and $(1, 0, 2\pi)$. (See Figure 6.8.)

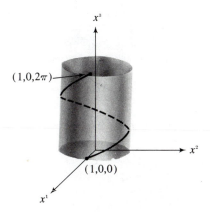

Figure 6.8   A helix.

Definition 5   *Let D be an open set in $\mathbb{R}^2$, with boundary S. A continuous function $f:(D \cup S) \rightarrow \mathbb{R}^n$ is called a (two-dimensional)* **surface** *in $\mathbb{R}^n$.*

Example 6

Let D be the rectangle $0 \leq \theta \leq 2\pi$, $0 \leq \phi \leq \pi$. Consider the function $f:(D \cup S) \rightarrow \mathbb{R}^3$ given by Equations (6.14). Notice that

$$(f^1)^2 + (f^2)^2 + (f^3)^2 \equiv 1,$$

so that all points "on" the surface f lie on the unit sphere in $\mathbb{R}^3$ (Figure 6.9).

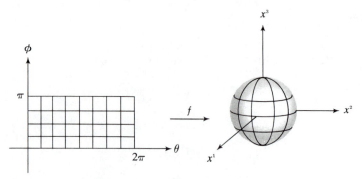

Figure 6.9   A spherical map.

Conversely, by trigonometry one can easily see that every point on the unit sphere is the image of some point (θ, ϕ) of $D \cup S$. The function f in fact could be used to construct a map of the globe (but most actual maps utilize more complicated functions than this). The coordinates (θ, ϕ) associated with a given point x, when expressed in degrees, are the longitude and colatitude of x, respectively.

Exercises

1.   Let $f(x^1, x^2) = |x^1 + x^2|$. Given $x_0 \in \mathbb{R}^2$ and $\varepsilon > 0$, determine $\delta > 0$ such that $|f(x) - f(x_0)| < \varepsilon$ for $\|x - x_0\| < \delta$.

2.   Let $f(t) = (t \cos t, \, t \sin t)$, so that $f : \mathbb{R}^1 \to \mathbb{R}^2$. Given $\varepsilon > 0$ determine $\delta > 0$ such that $\|f(t) - f(\pi)\| < \varepsilon$ provided $|t - \pi| < \delta$.

3.   Describe the following space curves geometrically:
   (a) $f(t) = (t \cos t, \, t \sin t, \, t)$ $(0 \le t \le 4\pi)$;
   (b) $f(t) = (\cos t, \, \sin t, \, \sin t)$ $(0 \le t \le 2\pi)$;
   (c) $f(t) = (t, 2t, 3t)$ $(-\infty < t < \infty)$.

4.   Describe the following surface geometrically:

$$f^1(\theta, \phi) = (2 + \cos \phi) \cos \theta,$$
$$f^2(\theta, \phi) = (2 + \cos \phi) \sin \theta,$$
$$f^3(\theta, \phi) = \sin \phi,$$

where $0 \le \theta, \, \phi \le 2\pi$. Sketch the surface and its "coordinate curves" $\theta = $ const., and $\phi = $ const.

5. If x_1, x_2 are given points of $\mathbb{R}^n$, find a curve $f : [0, 1] \to \mathbb{R}^n$ which corresponds to the straight line segment from x_1 to x_2.

6. Let $f = (f^1, f^2, \ldots, f^m)$ be a function from $D \subset \mathbb{R}^n$ to $\mathbb{R}^m$. Prove that f is continuous at $x_0 \in D$ if and only if each $f^j : D \to \mathbb{R}$, $(j = 1, 2, \ldots, m)$ is so.

7. Let $f : \mathbb{R}^n \to \mathbb{R}$ be a continuous function on $\mathbb{R}^n$. Show that for any constant c,

   (a) $\{x \in \mathbb{R}^n \mid f(x) < c\}$ is open,
   (b) $\{x \in \mathbb{R}^n \mid f(x) \le c\}$ is closed,
   (c) $\{x \in \mathbb{R}^n \mid f(x) = c\}$ is closed.

8. Describe the following sets geometrically, and determine whether they are open, closed, or neither.

   (a) $\{(x, y) \in \mathbb{R}^2 \mid xy > 1\}$,
   (b) $\{(x, y) \in \mathbb{R}^2 \mid x < y \le x + 1\}$,
   (c) $\{(x, y, z) \in \mathbb{R}^3 \mid x + y + z \le 1 \quad \text{and} \quad x \ge 0, y \ge 0, z \ge 0\}$.

9. Show that the image of a curve $f : [a, b] \to \mathbb{R}^n$ is a bounded set in $\mathbb{R}^n$. (See Exercise 6.)

10. Prove that "division is a continuous operation" with the obvious proviso.

6.4 *Properties of Continuous Functions*

In this section we generalize to n dimensions the theorem that a continuous function defined on a closed, bounded interval is bounded and assumes minimum and maximum values (cf. Section 4.5). We also consider uniform continuity and uniform convergence.

Definition   *A subset S of $\mathbb{R}^n$ which is both closed and bounded is said to be* **compact.**

Theorem 1   *A set $D \subset \mathbb{R}^n$ is compact if and only if every sequence in D contains a subsequence converging to a point of D.*

Proof Suppose first that D is compact. Then any sequence $\{x_j\}$ in D is bounded and must have a convergent subsequence $\{x_{j_k}\}$ by the Bolzano-Weierstrass theorem. Since D is closed, the limit of this subsequence is a point of D.

Conversely, suppose every sequence in D has a subsequence converging to some point of D. In particular, given $x_j \in D$, if $x_j \to x_0$ as $j \to \infty$, then $x_0 \in D$; this shows that D is closed. If D is unbounded, then there is a sequence $\{x_j\}$ in D with $\|x_j\| \to +\infty$. Clearly $\{x_j\}$ can have no convergent subsequence. Therefore D is also bounded. ∎

The next theorem generalizes the fact that a continuous function on a closed, bounded interval is bounded, to the case of a continuous function $f : D \to \mathbb{R}^m$.

Theorem 2   *Let D be a compact set in $\mathbb{R}^n$, and suppose $f : D \to \mathbb{R}^m$ is a continuous function. Then the image $f[D]$ is also a compact set.*

Proof Suppose first that $f[D]$ is not a bounded set. Then there must exist a sequence $\{x_j\}$ in D such that $\|f(x_j)\| \to \infty$ as $j \to \infty$. Let $\{x_{j_k}\}$ be a subsequence of $\{x_j\}$ converging to a point $x_0 \in D$. Since f is continuous at x_0, we have $f(x_{j_k}) \to f(x_0)$ as $k \to \infty$, so that

$$\|f(x_{j_k})\| \to \|f(x_0)\| \quad \text{as} \quad k \to \infty.$$

From this contradiction we conclude that $f[D]$ must be bounded.

To show that $f[D]$ is closed, we show that if $\{z_j\}$ is any sequence in $f[D]$, converging to some point z_0, then $z_0 \in f[D]$. To see this, note that $z_j = f(x_j)$ for some $x_j \in D$. The sequence $\{x_j\}$ has a subsequence $\{x_{j_k}\}$ with $\lim_{k \to \infty} x_{j_k} = x_0 \in D$. Hence, $z_{j_k} = f(x_{j_k}) \to f(x_0)$ as $k \to \infty$. But also $z_{j_k} \to z_0$ as $k \to \infty$, so that $z_0 = f(x_0)$, that is, $z_0 \in f[D]$. ∎

Corollary   *Let D be a compact set in $\mathbb{R}^n$ and let $f : D \to \mathbb{R}$ be continuous on D. Then $f(x)$ assumes both a maximum and a minimum value on D.*

Proof Since $f[D]$ is bounded by Theorem 2, $\sup_{x \in D} f(x) = M$ is finite. Let $\{f(x_j)\}$ be a sequence in $f[D]$ such that $f(x_j) \to M$ as $j \to \infty$. Since $f[D]$ is closed, we see that $M \in f[D]$. This shows that $M = \max_{x \in D} f(x)$. The proof for the minimum is similar. ∎

Let us recall the case of an ordinary, continuous function $f : [a, b] \to \mathbb{R}$. For locating points x_0 of maximum or minimum values of $f(x)$, there are three alternatives:

(1) x_0 is an end point a or b,   or

(2) $f'(x_0) = 0$,   or

(3) $f'(x_0)$ does not exist.

A similar situation prevails for functions of several variables; for simplicity we consider only the case $n = 2$.

Theorem 3   *Let f be a continuous real-valued function defined on a compact set $D \subset \mathbb{R}^2$. If the maximum (or minimum) value of f occurs at a point $(x_0, y_0) \in D$, then*

(1) (x_0, y_0) is a boundary point of D, or

(2) $\dfrac{\partial f}{\partial x}(x_0, y_0) = \dfrac{d}{dx} f(x, y_0)\Big|_{x=x_0} = 0$ and

$\dfrac{\partial f}{\partial y}(x_0, y_0) = \dfrac{d}{dy} f(x_0, y)\Big|_{y=y_0} = 0$,   or

(3) one or both of the above derivatives does not exist.†

Proof   Unravelling the logic, we see that what we must prove is: if (x_0, y_0) is an *interior point* of D (that is, $N_\varepsilon(x_0, y_0) \subset D$ for some $\varepsilon > 0$) *and* if the partial derivatives

$$\frac{\partial f}{\partial x}(x_0, y_0) \quad \text{and} \quad \frac{\partial f}{\partial y}(x_0, y_0)$$

exist, then these derivatives vanish. But this is fairly obvious; consider, for example, the function $f(x, y_0)$ (y_0 being fixed). This function of x is defined for $|x - x_0| < \varepsilon$ and has a maximum at $x = x_0$. Hence its derivative, which is precisely $\partial f/\partial x$, vanishes at (x_0, y_0). Similarly $\partial f/\partial y$ vanishes at (x_0, y_0). (See Figure 6.10.) ∎

---

† The expressions $\partial f/\partial x(x_0, y_0)$ and $\partial f/\partial y(x_0, y_0)$ are called the *partial derivatives* of f *with respect to* x and y, respectively, at the point (x_0, y_0). Note that $\partial f/\partial x$ is calculated simply by differentiating $f(x, y)$ with respect to x, treating y as a constant.

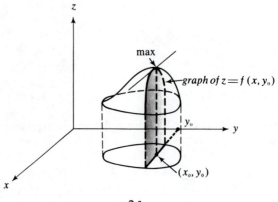

Figure 6.10 $\dfrac{\partial f}{\partial x}(x_0, y_0) = 0.$

Example

Locate the maximum and minimum of the function $f(x, y) = x^2 + y$ on the ellipse $x^2 + 4y^2 \leq 4$.

Solution   Note first that the inequality $x^2 + 4y^2 \leq 4$ determines a closed, bounded (that is, compact) set, so that f has a maximum and minimum.
Checking for vanishing of partial derivatives, we have

$$\frac{\partial f}{\partial x} = 2x \quad \text{and} \quad \frac{\partial f}{\partial y} = 1.$$

Since $\partial f/\partial y$ never vanishes, and since $\partial f/\partial x$ and $\partial f/\partial y$ exist everywhere, the only possibility is that the maximum and the minimum occur on the boundary curve $x^2 + 4y^2 = 4$.
Substituting $x^2 = 4 - 4y^2$, we are left with the problem of finding the maximum and minimum of the function†

$$g(y) = 4 - 4y^2 + y \quad (-1 \leq y \leq 1).$$

Solving this simple problem of ordinary calculus, we find that $g(y)$ has a maximum at $y = 1/8$, with $g(1/8) = 65/16$, and a minimum at the end point $y = -1$, with $g(-1) = -1$.

---

† Alternatively, the method of *Lagrange multipliers*, described in most calculus texts, can be used.

Therefore,

$$f_{\max} = f\left(\pm\frac{\sqrt{63}}{4}, \frac{1}{8}\right) = \frac{65}{16},$$

$$f_{\min} = f(0, -1) = -1.$$

We return briefly to theoretical considerations.

Theorem 4   *Let D be a compact set in $\mathbb{R}^n$ and let $f : D \to \mathbb{R}^m$ be continuous on D. Then f is uniformly continuous on D.*

Theorem 5   *Let D be an arbitrary set in $\mathbb{R}^n$ and let $\{f_j\}$ be a sequence of functions from D into $\mathbb{R}^m$. Then the sequence $\{f_j\}$ converges uniformly to some function $f : D \to \mathbb{R}^m$ if and only if the sequence $\{f_j\}$ satisfies the uniform Cauchy condition on D, viz.*

$$d_D(f_j, f_k) = \sup_{x \in D} \|f_j(x) - f_k(x)\| \to 0 \quad \text{as} \quad j, k \to \infty.$$

Theorem 6   *Let D and $\{f_j\}$ be as in Theorem 5. If each function f_j is continuous on D, and if $f_j \to f$ uniformly on D, then f is also continuous on D.*

The proofs of these three theorems (as well as the required definitions of uniform continuity and so forth) are virtually identical to the corresponding proofs in Chapter 4. You are asked to locate these proofs and carry out the required modifications.

Exercises

1. Locate the maximum and minimum values of

    (a) $f(x, y) = 3x - y$ on the square $-1 \le x, y \le 1$.
    (b) $f(x, y) = xy$ on the disk $x^2 + y^2 \le 1$.
    (c) $f(x, y) = xye^{-x^2-y^2}$ on the first quadrant $x, y \ge 0$.
    (d) $f(x, y) = x^2 + y^2$ on the straight line $x + 2y = 1$.

2. Show that a curve joining two points in $\mathbb{R}^n$ is a compact set. (Literally speaking, this is false; correct this statement and then prove it.)

3. Let D be a compact set and let $x_0 \notin D$. Define $f:D \to \mathbb{R}$ by $f(x) = \|x - x_0\|$. Show that $\min_{x \in D} f(x)$ exists and is positive.

   The number $\rho(x_0, D) = \min_{x \in D} \|x - x_0\|$ is called the *distance from* x_0 *to* D.

4. Let $\rho(x_0, D)$ be as in Exercise 3. Show that

   (a) $\rho(x_0, D) \leq \|x_0 - x_1\| + \rho(x_1, D)$; and therefore,

   (b) $\rho(x_0, D)$ is a uniformly continuous function of x_0.

5. Let $f: D \to \mathbb{R}^m$, where D is a closed *unbounded* set in $\mathbb{R}^n$, and f is continuous on D. Assuming that $f(x) \to 0$ as $\|x\| \to \infty$ $(x \in D)$, prove that f is uniformly continuous on D. (*Hint:* Given $\varepsilon > 0$, first pick M such that $\|f(x)\| < \varepsilon/2$ for $\|x\| \geq M$, $x \in D$.)

6. Let D be a closed, unbounded set in $\mathbb{R}^n$, and let f be a continuous, non-negative real-valued function on D. Assuming that $f(x) \to 0$ as $\|x\| \to \infty$ $(x \in D)$ and that $f(x_0) > 0$ for some point $x_0 \in D$, show that f assumes a maximum on D. Does f necessarily assume a minimum on D?

7. Let $\{f_j\}$ be a sequence of functions from $D \subset \mathbb{R}^n$ to $\mathbb{R}^m$ satisfying

   $$\|f_{j+1}(x) - f_j(x)\| \leq a_j \quad \text{for } x \in D \text{ and } j = 1, 2, 3, \ldots.$$

   Show that if the series $\sum_1^\infty a_k$ converges then $\{f_j\}$ satisfies the uniform Cauchy condition on D, and hence converges uniformly.

6.5 *An Existence Theorem for Ordinary Differential Equations*

As an example of the applications of the theory developed in this text we will in this section prove a fundamental theorem pertaining to the existence of solutions to (nonlinear) systems of ordinary differential equations. This theorem is often referred to, without proof, in introductory courses in differential equations.

Consider the (vector) equation

$$\frac{dx}{dt} = f(x, t) \tag{6.16}$$

where $x = x(t)$ represents a function $x: \mathbb{R} \to \mathbb{R}^n$ (i.e. $x = x(t)$ is a *curve* in *n*-space), and where $f: \mathbb{R}^{n+1} \to \mathbb{R}^n$ is a given function. Equation (6.16) is called a *first-order system of ordinary differential equations*; it can be written out as:

$$\left. \begin{array}{c} \dfrac{dx^1}{dt} = f^1(x^1, x^2, \ldots, x^n, t) \\[1em] \cdot \\ \cdot \\ \cdot \\[1em] \dfrac{dx^n}{dt} = f^n(x^1, x^2, \ldots, x^n, t) \end{array} \right\} \qquad (6.17)$$

A differentiable function $x(t)$ is called a *solution* of the system (6.16), for $a < t < b$, if it satisfies the equation identically as a function of t:

$$\frac{dx(t)}{dt} \equiv f(x(t), t), \quad \text{for } a < t < b.$$

The *initial value problem* is the problem of finding a solution to the system (6.16) that also satisfies given *initial conditions*

$$x(a) = x_0 \qquad (6.18)$$

where $x_0 \in \mathbb{R}^n$ is given.

We will henceforth assume that the function f in Equation (6.16) is defined on a domain $D \subset \mathbb{R}^{n+1}$ and satisfies the following:

Lipschitz condition: There exists a constant K such that

$$\|f(x, t) - f(y, t)\| \le K\|x - y\| \quad \text{for all } (x, t), (y, t) \in D. \qquad (6.19)$$

This condition holds, for example, if D is compact and f is continuously differentiable on D; we will not prove this here (or even define differentiability in *n* dimensions).

Under the above condition it can be proven that the *initial value* problem (6.16), (6.18) *has a unique solution* $x(t)$. The proof consists of two parts: (a) uniqueness, and (b) existence. The following simple lemma is needed for the uniqueness proof.

Lemma 1   Let σ be a differentiable function satisfying

$$\sigma'(t) \le A\sigma(t) \quad \text{for } a \le t \le b \tag{6.20}$$

where $A = $ constant. Then

$$\sigma(t) \le \sigma(a)e^{A(t-a)} \quad \text{for } a \le t \le b. \tag{6.21}$$

Proof   From (6.20) we have

$$\frac{d}{dt}\{e^{-At}\sigma(t)\} = e^{-At}\{\sigma'(t) - A\sigma(t)\} \le 0.$$

Hence $e^{-At}\sigma(t)$ is nonincreasing on $[a, b]$, and (6.21) follows.   ∎

Theorem 1 (Uniqueness theorem)   Suppose f satisfies the Lipschitz condition (6.19). Then there is at most one solution $x(t)$ to the initial value problem (6.16), (6.18).

Proof   Suppose $x(t)$ and $y(t)$ are solutions. Define

$$\sigma(t) = \|x(t) - y(t)\|^2 = (x - y) \cdot (x - y). \tag{6.22}$$

Then from (6.16) we obtain

$$\begin{aligned}
\sigma'(t) &= 2(x - y) \cdot (x' - y') \\
&= 2(x - y) \cdot (f(x, t) - f(y, t)) \\
&\le 2\|x - y\| \, \|f(x, t) - f(y, t)\| \quad \text{(by Schwarz's inequality)} \\
&\le 2K\|x - y\|^2. \qquad\qquad\qquad \text{(by (6.19))}
\end{aligned}$$

Thus $\sigma'(t) \le 2K\sigma(t)$. By Lemma 1, therefore,

$$\sigma(t) \le \sigma(a)e^{2K(t-a)}.$$

But $\sigma(a) = 0$ because $x(a) = y(a) = x_0$ by (6.18). Since $\sigma(t) \ge 0$ by its definition (6.22), we conclude that $\sigma(t) \equiv 0$, i.e. $x(t) \equiv y(t)$.   ∎

We turn now to the existence theorem, which is somewhat more difficult. The proof of existence consists of replacing the initial value problem by an equivalent *integral equation* (Lemma 2), which is then solved by a method of *iteration* (Theorem 2).

Lemma 2   *The initial value problem* (6.16), (6.18) *is equivalent to the integral equation*

$$x(t) = x_0 + \int_a^t f(x(s), s) \, ds, \tag{6.23}$$

in the sense that any solution $x(t)$ of the initial value problem is also a solution of (6.23), *and conversely.*

Proof   First, suppose $x(t)$ satisfies (6.23). Then $x(a) = x_0$, and (by Theorem 1, Section 5.3)

$$\frac{dx}{dt} = f(x(t), t),$$

that is, $x(t)$ solves the initial value problem. The converse follows equally easily (from Theorem 3, Section 5.3). ∎

The method of iteration (which was discussed in a much simpler setting in Chapter 1) consists of choosing an initial "guess" $x(t) = x_1(t)$ to the solution of Equation (6.23), and then "iterating," i.e. defining a sequence of functions $\{x_n(t)\}$ inductively by:

$$x_{n+1}(t) = x_0 + \int_a^t f(x_n(s), s) \, ds. \tag{6.24}$$

If we can prove that $x_n \to x$ in a suitable way (viz. *uniformly*), then it will follow from (6.24) that $x(t)$ satisfies (6.23), and this establishes the existence of a solution to the initial value problem.

Theorem 2 (Existence theorem)   *Suppose f satisfies the Lipschitz condition* (6.19) *for all t such that $|t - a| \le h$, and for all x, $y \in \mathbb{R}^n$. Then there exists a differentiable function $x(t)$ defined for $|t - a| \le h$ and satisfying* (6.23).

Proof   Choose $x_1(t) \equiv x_0$, and define $x_n(t)$ for $n \ge 2$ by Equation (6.24). It follows by induction that $x_n(t)$ is defined and continuously differentiable on $|t - a| \le h$.

  Let

$$M = \max_{t \in [a-h, a+h]} \| f(x_0, t) \|; \tag{6.25}$$

M exists finitely by the Corollary to Theorem 2 of Section 6.4. We then have, for $|t - a| \le h$

$$\|x_2(t) - x_1(t)\| = \left\| \int_a^t f(x_0, s) \, ds \right\| \le \left| \int_a^t \|f(x_0, s)\| \, ds \right| \le M|t - a| \quad (6.26)$$

(proof of the first inequality being left for the Exercises at the end of this section).
   Similarly,

$$\|x_3(t) - x_2(t)\| \le \left| \int_a^t \|f(x_2(s), s) - f(x_1(s), s)\| \, ds \right|$$

$$\le K \left| \int_a^t \|(x_2(s) - x_1(s)\| \, ds \right| \quad \text{(by (6.19))}$$

$$\le KM \left| \int_a^t |s - a| \, ds \right| \quad \text{(by (6.26))}$$

$$= \frac{KM}{2} |t - a|^2.$$

Continuing inductively, we obtain

$$\|x_{n+1}(t) - x_n(t)\| \le \frac{K^{n-1}M}{n!} |t - a|^n \le \frac{K^{n-1}Mh^n}{n!} \quad (6.27)$$

for all n. Since the series $\sum_0^\infty (Kh)^n/n!$ converges it follows (see Exercise 7 of the previous section) that $\{x_n(t)\}$ satisfies the uniform Cauchy condition on $|t - a| \le h$, and hence converges uniformly to a continuous function $x(t)$. Since also $\{f(x_n(t), t)\}$ converges uniformly to $f(x(t), t)$ on $|t - a| \le h$ (why?), Equation (6.24) implies that

$$x(t) = x_0 + \int_a^t f(x(s), s) \, ds. \quad \blacksquare$$

Exercises

1. Solve the initial value problem

$$\frac{dx}{dt} = ax, \quad x(0) = 1$$

by the method of iteration. (You will generate the Taylor expansion of $x(t) = e^{at}$.)

2. Explain why the existence theorem proved in this section does not apply to the problem

$$\frac{dx}{dt} = x^2, \quad x(0) = 1.$$

(A more complicated version of our proof can be given, which will show that equations of this kind always have solutions for $|t - a|$ sufficiently small.†)

3. If f, g are integrable real-valued functions on $[a, b]$ show that

$$\left| \int_a^b f(x)g(x)\, dx \right| \le \left(\int_a^b |f(x)|^2\, dx \right)^{1/2} \left(\int_a^b |g(x)|^2\, dx \right)^{1/2}$$

(Schwarz's inequality for integrals: see the proof of Schwarz's inequality given on page 190).

4. Let $f: [a, b) \to \mathbb{R}^n$ and suppose f is integrable on $[a, b]$ in the sense that each component f^i is. Show that

$$\left\| \int_a^b f(x)\, dx \right\| \le \int_a^b \| f(x) \|\, dx$$

(note that $\int_a^b f(x)\, dx$ is an n-vector).

---

† See, for example, G. Birkhoff and G.-C. Rota, *Ordinary Differential Equations*, Blaisdell (1969), Sec. 6.9.

Appendix

I Logic

Although mathematics is usually thought to be a completely logical subject, many students of mathematics are quite uncertain about the basic principles of logic. As used in mathematics, logic has two parts: the "statement calculus," which deals with logical connectives, such as "and," "or," "not," and "implies"; and the "quantifier calculus," which deals with the quantifying phrases "for some" ("there exists") and "for all."

The proper use of quantifiers is especially important in the study of mathematical analysis. The definition of limit, for example, uses *three* quantifying phrases:" For every $\varepsilon > 0$, there exists $\delta > 0$, such that for every $x, \ldots$." Sentences of this degree of logical complexity are seldom if ever encountered in a nonmathematical setting. The purpose of this Appendix is to introduce you to the most frequently used rules of statement and quantifier logic.

I.1 Logical Connectives

The study of logic is concerned with the *truth* or *falseness* of statements. For example,

$$2 + 2 = 4$$

is a true statement (as could be proved from the axioms and definitions of arithmetic). The statement

$$x^2 < 1,$$

however, is neither true nor false, since it contains an unspecified *variable*, *x*. In this section we consider only statements without variables.

Let *A* be a given statement. Then we can form the *negation* of *A*, which we denote by

$$\sim A = \text{not } A.$$

If *A* is true, then $\sim A$ is false, and moreover, if *A* is false, then $\sim A$ is true.

The operation of negation can be described by the following simple "*truth table.*"

A	$\sim A$
T	F
F	T

Given statements *A* and *B*, we can form the *conjunction*

$$A \,\&\, B = A \text{ and } B.$$

Whether the statement *A* & *B* is true or false depends on the truth of falsity of *A* and *B*. We define *A* & *B* to be a true statement if and only if both *A* and *B* are true. The truth table for *A* & *B* is, therefore, the following.†

A	*B*	*A* & *B*
T	T	T
F	T	F
T	F	F
F	F	F

Another common logical connective is the *disjunction*

$$A \vee B = A \text{ or } B.$$

---

† The arrangement of the truth columns for *A* and *B* in this table is standard and is used in all truth tables. It can easily be expanded to construct truth tables with several statements *A, B, C, ...*

In mathematical logic, this is always the *inclusive* disjunction, so that $A \vee B$ is true whenever either A is true or B is true, or both: †

A	B	$A \vee B$
T	T	T
F	T	T
T	F	T
F	F	F

Next we have logical *implication*:

$$A \Rightarrow B \quad = \quad A \quad \text{implies} \quad B \quad = \quad \text{if } A \text{ then } B.$$

The truth table for implication is

A	B	$A \Rightarrow B$
T	T	T
F	T	T
T	F	F
F	F	T

Some readers may disagree with this table at first, especially with the last row, which says that "false implies false is true." Thus the statement

$$3 + 1 = 7 \Rightarrow 6 - 1 = 2$$

is a true statement! (In fact we can prove it! Suppose $3 + 1 = 7$. Then

$$3 + 1 - 3 = 7 - 3 = 4, \quad \text{so} \quad 1 = 4.$$

Therefore $6 - 1 = 6 - 4 = 2$.)

---

† Everyday language uses both inclusive and exclusive disjunction, so that ambiguity can arise. To avoid ambiguity, some authors are reduced to using the phrase "and/or" to indicate inclusive disjunctions.

The logical connective *if and only if* :

$$A \Leftrightarrow B \quad = \quad A \quad \text{if and only if} \quad B$$

has the truth table

A	B	$A \Leftrightarrow B$
T	T	T
F	T	F
T	F	F
F	F	T

A statement *P* constructed from various substatements $A, B, C, \ldots$, and which is true no matter whether $A, B, C, \ldots$ are true or false, is called a *tautology*.

Example 1

$[A \,\&\, (A \Rightarrow B)] \Rightarrow B$ is a tautology.

(This says that if *A* is true and $A \Rightarrow B$, then *B* is true.)

To verify this, we construct the following truth table, using the given truth tables for & and $\Rightarrow$.

A	B	$A \Rightarrow B$	$A \,\&\, (A \Rightarrow B)$	$[A \,\&\, (A \Rightarrow B)] \Rightarrow B$
T	T	T	T	T
F	T	T	F	T
T	F	F	F	T
F	F	T	F	T

Since the final column contains only "T's," the given assertion is a tautology.†

---

† Note that this verification is completely routine. In fact, it is easy to program a computer to verify tautologies.

A statement P which is false under all circumstances is called a *contradiction.* An example: $P = (A \,\&\, {\sim}A)$. Clearly P is a contradiction if and only if ${\sim}P$ is a tautology. Contradictions are often employed as part of a proof. Thus if P is a statement ("theorem") to be proved, then proving that

$$\sim P \Rightarrow Q$$

where Q is a contradiction, establishes the truth of P.

Two statements P and Q, constructed from the same substatements A, B, $C, \dots$, are said to be *logically equivalent* if they have identical truth tables. In this case we write $P \equiv Q$.

Example 2

${\sim}(A \vee B) \equiv ({\sim}A) \,\&\, ({\sim}B)$.   (In words, $A \vee B$ is false if and only if A and B are both false.) To verify this we again construct a truth table.

A	B	$A \vee B$	${\sim}(A \vee B)$	${\sim}A$	${\sim}B$	$({\sim}A) \,\&\, ({\sim}B)$
T	T	T	F	F	F	F
F	T	T	F	T	F	F
T	F	T	F	F	T	F
F	F	F	T	T	T	T

Since the columns for ${\sim}(A \vee B)$ and $({\sim}A) \,\&\, ({\sim}B)$ are identical, we have proved the equivalence of these statements.

The foregoing logical operations are basic ingredients of all mathematics that every student of the subject should be familiar with. The following exercises cover many of the standard uses of logical connectives encountered in mathematics.

Exercises

1. Verify the following equivalences by means of truth tables.

    (a) $A \Rightarrow B \equiv ({\sim}B) \Rightarrow ({\sim}A)$.

    (*Note*: $({\sim}B) \Rightarrow ({\sim}A)$ is called the *contrapositive* of $A \Rightarrow B$.)

    (b) $A \vee B \equiv ({\sim}A) \Rightarrow B$.

(c) $\sim(A \,\&\, B) \equiv (\sim A) \vee (\sim B)$.

(This should be compared with Example 2.)

(d) $\sim(A \Rightarrow B) \equiv A \,\&\, (\sim B)$.

(e) $A \vee B \equiv B \vee A$.

(f) $A \Leftrightarrow B \equiv (A \Rightarrow B) \,\&\, (B \Rightarrow A)$.

(g) $A \vee (B \,\&\, C) \equiv (A \vee B) \,\&\, (A \vee C)$.

(Note that the truth table for 3 statements A, B, C must contain $2^3 = 8$ rows.)

(h) $A \,\&\, (B \vee C) \equiv (A \,\&\, B) \vee (A \,\&\, C)$.

(i) $A \Rightarrow (B \Rightarrow C) \equiv (A \,\&\, B) \Rightarrow C$.

(j) $(A \vee B) \Rightarrow C \equiv (A \Rightarrow C) \,\&\, (B \Rightarrow C)$.

(k) $A \Rightarrow (B \,\&\, C) \equiv (A \Rightarrow B) \,\&\, (A \Rightarrow C)$.

2. Show that the following statements are tautologies.

(a) $A \Rightarrow A$.

(b) $\sim(A \,\&\, \sim A)$ (Law of excluded middle).

(c) $[(A \Rightarrow B) \,\&\, (B \Rightarrow C)] \Rightarrow (A \Rightarrow C)$ (Law of syllogism).

(d) $[A \Rightarrow (B \,\&\, \sim B)] \Rightarrow \sim A$ (Reductio ad absurdum).

3. If $P \Rightarrow Q$ is a tautology, we say that P is *logically stronger* than Q. Which (if any) of the following statements is logically stronger?

(a) $(A \,\&\, B) \Rightarrow C$; $(A \Rightarrow C) \,\&\, (B \Rightarrow C)$.

(b) $A \Rightarrow (B \Rightarrow C)$; $(A \Rightarrow B) \Rightarrow C$.

4. The statement $B \Rightarrow A$ is called the *converse* of the statement $A \Rightarrow B$. Show that a statement and its converse are not logically comparable.

5. Show that the logical connectives of implication and conjunction (for example) could both be dispensed with in logic.

6. Give truth tables for each of the following assertions.

(a) either A or B but not both.

(b) A is true unless B is true.

(c) A is a necessary condition for B.

(d) A is a sufficient condition for B.

(e) A and B are independent.

7. Express the assertions of Exercise 6 in terms of $\&$, $\vee$, $\sim$, $\Rightarrow$.

8. Given that $3 + 1 = 7$, "prove" that $1 = 0$, and hence that all numbers are equal.

I.2   Quantifiers

Let S denote a given set; $x \in S$ means that x is a *member* of S. Let $A(x)$ denote an assertion about x; for each choice of x, the assertion $A(x)$ is either true or false.

We can consider the statement

I.   For every $x \in S$, $A(x)$ is true, which we notate as

$$\text{I.} \quad \forall \, x \in S, \, A(x).$$

The phrase "for all," or "for every," ($\forall$) is called the *universal quantifier*. The statement $\forall \, x \in S, \, A(x)$ is said to be *true* provided, naturally, that $A(x)$ is true for every choice of x in S. If S is a finite set, say

$$S = \{x_1, x_2, \ldots, x_n\},$$

then $\forall \, x \in S, \, A(x)$ means the same as

$$A(x_1) \,\&\, A(x_2) \,\&\, \cdots \,\&\, A(x_n).$$

If S is an infinite set, there is no such simple interpretation.

We are also interested in the statement

II.   For some $x \in S$, $A(x)$ is true, which we abbreviate as

$$\text{II.} \quad \exists \, x \in S, \, A(x).$$

The phrase "for some," or "there exists," ($\exists$) is called the *existential quantifier*. The statement $\exists \, x \in S, \, A(x)$ is *true* provided that $A(x)$ is a true statement for at least one x in S. If S is finite,

$$S = \{x_1, x_2, \ldots, x_n\},$$

then $\exists \, x \in S, \, A(x)$ means the same as

$$A(x_1) \lor A(x_2) \lor \cdots \lor A(x_n).$$

The universal and existential quantifiers are thus seen to be extensions of the logical connectives $\&$ and $\lor$, respectively, to deal with infinitely many assertions, or assertions about infinitely many "things" x.

Example 1

Let $\mathbb{R}$ denote the set of all real numbers, and $\mathbb{Q}$ the set of all rational numbers. We ask the reader to decide on the basis of his experience in mathematics, whether the following statements are true or false.

(a) $\forall\, x \in \mathbb{R}$, $2x > x$ T or F?

(b) $\exists\, x \in \mathbb{Q}$, $x^2 = 2$ T or F?

(c) $\forall\, x \in \mathbb{Q}$, $x^2 \in \mathbb{Q}$,   T or F?

(d) $\exists\, x \in \mathbb{R}$, $x^5 - 3x^2 + 5 = 0$ T or F?

Note that (b) and (d) are not completely trivial!

The symbol "x" in the statements I, II is called a *quantified variable*. There are two basic regulations regarding the use of variables in mathematics. The first states that any variable which occurs *must be quantified*. For example,

$$x > 2$$

makes no sense by itself, but only makes sense if the meaning of x is specified. Another example:

$$x > 2 \Rightarrow x^2 > 4;$$

this seems to be correct, because we are conditioned to interpret it as:

$$\forall\, x \in \mathbb{R}, \; x > 2 \Rightarrow x^2 > 4. \tag{1}$$

The latter statement makes sense and is true. Unwritten quantifiers of this kind occur frequently in mathematical writing; they are acceptable provided the context is absolutely clear.

The second rule is that a quantified variable is a "dummy" variable and can be replaced (in all its occurrences) by any other variable symbol. For example,

$$\forall\, t \in \mathbb{R}, \, t > 2 \Rightarrow t^2 > 4$$

carries precisely the same meaning as Statement (1) above.

It is often useful to know how to *negate* a quantified statement. For this we have the rules

(i) $\sim(\forall\, x \in S, A(x)) \equiv \exists\, x \in S, \sim A(x),$

(ii) $\sim(\exists\, x \in S, A(x)) \equiv \forall\, x \in S, \sim A(x).$

Rule (i) is the rule of "*counterexample.*" Rule (ii) can be derived from Rule (i). Note that if S is a finite set, rules (i) and (ii) are just known tautologies; for example, suppose $S = \{x_1, x_2\}$. Then

$$\sim(\forall\, x \in S,\, A(x)) \equiv \sim(A(x_1) \,\&\, A(x_2))$$

$$\equiv (\sim A(x_1)) \vee (\sim A(x_2)) \equiv \exists\, x \in S,\, \sim A(x).$$

Other combinations of quantifiers and connectives will be found in the Exercises.

In mathematical analysis it is common to encounter statements involving several quantifiers. For example, the definition of the limit of a function uses three quantifiers: $\lim_{x \to a} f(x) = L$ is defined to mean

$$\forall\, \varepsilon > 0,\, \exists\, \delta > 0,\, \forall\, x \in \mathbb{R},\, (0 < |x - a| < \delta \Rightarrow |f(x) - L| < \varepsilon). \quad (2)$$

In handling such statements, it is essential to understand first *that the order in which quantifiers appear affects the meaning of the statement.*

Example 2

Consider the statements

(a) $\qquad\qquad\qquad \forall\, x \in \mathbb{R},\quad \exists\, y \in \mathbb{R},\quad x + y = 1,$

(b) $\qquad\qquad\qquad \exists\, y \in \mathbb{R},\quad \forall\, x \in \mathbb{R},\quad x + y = 1.$

Statement (a) is true: given $x \in \mathbb{R}$, we can set $y = 1 - x$, and then $x + y = 1$. Statement (b), however, is false: there does *not* exist a $y \in \mathbb{R}$ such that $x + y = 1$ holds for all $x \in \mathbb{R}$.

To summarize, in any statement

$$\forall\, x \in S,\quad \exists\, y \in T,\quad A(x, y),$$

the choice of y is allowed to depend on the value of x, whereas in the statement

$$\exists\, y \in T,\quad \forall\, x \in S,\quad A(x, y),$$

the choice of y must be independent of x.

More generally, we can state the following rule: if an existential expression $\exists\, x \in T$ occurs in a statement, then the choice of x may depend on any variable that occurs *before* x in the statement, but not on any variable that occurs after x.

The negation of a statement with several quantifiers can be found by repeated application of Rules (i) and (ii) above.

Example 3

Consider the (false) statement (b) of Example 2. The negation of this statement is:

$$\sim(\exists \, y \in \mathbb{R}, \; \forall \, x \in \mathbb{R}, \; x + y = 1) \equiv \forall \, y \in \mathbb{R}, \; \sim(\forall \, x \in \mathbb{R}, \; x + y = 1)$$

$$\equiv \forall \, y \in \mathbb{R}, \; \exists \, x \in \mathbb{R}, \; x + y \neq 1.$$

(We could prove that this negated statement is true as follows: let $y \in \mathbb{R}$. Take $x = -y$; then $x + y = 0 \neq 1$.)

Exercises

1. Determine which of the following statements are true. ($\mathbb{R}$ denotes the set of real numbers, $\mathbb{N}$ the set of all integers.)

    (a) $\forall \, x \in \mathbb{R}, \; \exists \, y \in \mathbb{R}, \; y^2 \geq x$;
    (b) $\exists \, x \in \mathbb{R}, \; \forall \, y \in \mathbb{R}, \; y^2 \geq x$;
    (c) $\exists \, y \in \mathbb{R}, \; \forall \, x \in \mathbb{R}, \; y^2 \geq x$;
    (d) $\forall \, x \in \mathbb{R}, \; \exists \, y \in \mathbb{N}, \; x \leq y$;
    (e) $\forall \, x \in \mathbb{R}, \; \forall \, y \in \mathbb{R}, \; xy = yx$;
    (f) $\forall \, \varepsilon > 0, \exists \, \delta > 0, \forall \, x \in \mathbb{R}, \; |x| < \delta \Rightarrow x^2 < \varepsilon$.

2. Consider the assertion "every real number is smaller than some integer." Express this in formal notation, and decide whether it is true or false.

3. Let S be a given set. Decide (intuitively) which of the following assertions are valid.

    (a) $[\forall \, x \in S, p(x)] \Rightarrow [\exists \, x \in S, p(x)]$.
    (b) $[\forall \, x \in S, p(x)] \, \& \, [x_0 \in S] \Rightarrow p(x_0)$.
    (c) $\{[\forall x \in S, p(x)] \, \& \sim [x_0 \in S]\} \Rightarrow \sim p(x_0)$.
    (d) $[\forall \, x \in S, p(x) \lor q(x)] \Rightarrow [\forall \, x \in S, p(x)] \lor [\forall \, x \in S, q(x)]$.
    (e) Same as (d), with $\lor$ replaced by $\&$.
    (f) $[\forall \, x \in S, p(x)] \lor [\forall \, x \in S, q(x)] \Rightarrow \forall \, x \in S, p(x) \lor q(x)$.
    (g) $[\forall \, x \in S, p(x)] \, \& \, [\exists \, x \in S, p(x) \Rightarrow q(x)] \Rightarrow [\exists \, x \in S, q(x)]$.

4. Negate the following statements; decide whether each given statement or its negation is true.

(a) $\forall\, x \in \mathbb{N},\quad \exists\, y \in \mathbb{N},\quad y^2 = x.$

(b) $\forall\, x \in \mathbb{N},\quad \exists\, y \in \mathbb{N},\quad y = x^2.$

(c) $\forall\, x \in \mathbb{R},\quad \exists\, y \in \mathbb{R},\quad xy = 1.$

(d) $\exists\, x \in \mathbb{R},\quad \forall\, y \in \mathbb{R},\quad xy = 0.$

(e) $\forall\, x \in \mathbb{R},\quad \forall\, y \in \mathbb{R},\quad xy \neq 0 \Rightarrow \dfrac{1}{x} + \dfrac{1}{y} = \dfrac{1}{x+y}.$

(f) $\exists\, C \in \mathbb{R},\quad \forall\, x \in \mathbb{R},\quad (x \geq C \Rightarrow \exists\, y \in \mathbb{R},\ y^2 + y = x).$

I.3 Proof

Mathematics is a deductive science. Its essence lies in the proofs of theorems. This is not to underestimate the importance of intuition and insight in mathematics. The final product of mathematical creation, however, must be a rigorous logical theory.

What is a proof? This question frequently troubles beginning students of serious mathematics. (For example, can you prove that $2 + 2 = 4$? This fact is certainly not an axiom or a definition, so it must be a theorem.) Let us consider first an assertion of the form $A \Rightarrow B$, where A and B are given statements. By a (formal) *proof* of the assertion $A \Rightarrow B$, we mean a finite list of statements

$$A_1, A_2, A_3, \ldots, A_n$$

with the properties

(1) $A_1 = A$ (the *hypothesis*) and $A_n = B$ (the *conclusion*);
(2) Each statement A_i is either an axiom or a logical consequence of the previous statements $A_1, A_2, \ldots, A_{i-1}$.

Needless to say, "proofs" satisfying this definition are seldom encountered in practice. This is because we have invented several ways of simplifying the writing of proofs, such as (a) introducing definitions; (b) using known theorems without reproving them; (c) omitting steps in the proof which the reader is expected to fill in. We only understand the proof of a certain theorem when we are convinced that we could (upon demand) write out a complete formal proof of it.

Consider next the case of a quantified statement of the form $\forall\, x \in S, p(x)$. To prove this assertion, we prove instead the assertion $x \in S \Rightarrow p(x)$, according to the definition of proof just given.

Example

Theorem $\forall x \in (0, 1), x^2 < x.$

Proof: 1. Let $x \in (0, 1)$.   (hypothesis)

    2. Then $x > 0$ & $x < 1$.   (definition)

    3. Therefore, $x \cdot x < x \cdot 1$.   (axiom)

    4. But $x \cdot x = x^2$.   (definition)

    5. And $x \cdot 1 = x$.   (axiom)

    6. Therefore $x^2 < x$.   (substitution)

Finally, consider an existentially quantified statement $\exists\, x \in S$, $p(x)$. To prove this *directly* is to exhibit a member $x \in S$ for which $p(x)$ is true. For example, the statement $\exists\, x \in \mathbb{R}$, $x^2 = 2$ is proved in Chapter 1 by defining $x = \lim_{n \to \infty} x_n$ for a certain sequence $\{x_n\}$.

We can also prove the existential assertion $\exists\, x \in S$, $p(x)$ by the method of contradiction. Thus we would show that the negation, namely, $\forall\, x \in S$, $\sim p(x)$, leads to a contradiction. According to the law of the excluded middle $(\sim \sim A \equiv A)$, this is a valid proof. (Some mathematicians do not feel satisfied with such proofs of existence by contradiction, because proofs of this kind are sometimes "nonconstructive," in the sense that no method of actually determining a value of x for which $p(x)$ is true can be deduced from the proof. From a practical point of view, this is certainly a valid criticism.†)

This completes our discussion of mathematical logic. It should be stated that we have greatly oversimplified the subject.‡ Nevertheless, the principles discussed here, if combined with one's logical intuition, should suffice for most contemporary mathematics.

Exercises

Prove the following theorems. At each step indicate whether the statement is an axiom, definition, known theorem, or a logical step. Identify all tautologies used in the logical steps.

1. $\forall\, n \in \mathbb{N}$, n odd $\Rightarrow n^2$ odd.

2. $\forall\, x \in \mathbb{R}$, $|x|^2 = x^2$. (Use the definition $|x| = +x$ if $x \geq 0$ and $|x| = -x$ if $x < 0$. Consider cases.)

---

† The definitive work on this topic is E. Bishop, *Foundations of Constructive Analysis*, McGraw-Hill (1967).

‡ For further details see R. R. Stoll, *Set Theory and Logic*, W. H. Freeman (1961); or A. Tarski, *Introduction to Logic*, Oxford University Press (1941).

3. $\sim(\exists\, x \in \mathbb{R},\ x^2 = -1)$.

4. $\exists\, n \in \mathbb{N},\ 2^n > n^{10}$ and $n > 1$.

I.4 Sets and Functions

The most basic notions in mathematics, at least by contemporary standards, are the concepts of a *set* and of *membership* in a set. For example, the entire real number system can be built up by successive set formations, beginning only with the single number zero. Further set formation leads to n-dimensional spaces, functions, and so on, to all of calculus and analysis. We have no intention of carrying out such a program here; rather we will simply discuss the most elementary and useful properties of sets.

If X is a given set and $p(x)$ an assertion which applies to any member x of X, we use the notation

$$\{x \in X \mid p(x)\}$$

to denote the set of all members of X for which $p(x)$ is true. For example, $\{x \in \mathbb{R} \mid x^2 < x\}$ (read: the set of all x in $\mathbb{R}$ such that $x^2 < x$) can easily be seen to be equal to the interval

$$(0, 1) = \{x \in \mathbb{R} \mid 0 < x < 1\}.$$

Let $A, B, \ldots$ denote subsets of a given "universal" set X; in any given situation the set X would be clearly defined. We can then perform the following set operations:

1. *Union:* $A \cup B = \{x \in X \mid (x \in A) \vee (x \in B)\}$.
2. *Intersection:* $A \cap B = \{x \in X \mid (x \in A)\ \&\ (x \in B)\}$.
3. *Complement:* $A' = \{x \in X \mid \sim(x \in A)\}$.

Note that these operations correspond to the basic logical connectives "or," "and," and "not." The connective "implies" also has a role in set theory, as follows.

Definition 1 *We say that $A \subset B$ (A is **contained in** B) provided that*

$$\forall\, x \in X,\ (x \in A) \Rightarrow (x \in B).$$

Furthermore, we define $A = B$ to mean that $A \subset B$ and $B \subset A$; in other words, $A = B$ if and only if

$$\forall\, x \in X,\ (x \in A) \Leftrightarrow (x \in B).$$

The following useful set identities are simply set-theoretic reflections of various logical tautologies discussed in Section I.1.

(i) $(A')' = A$.

(ii) $A \cup A' = X$,

$A \cap A' = \emptyset$ (the empty set).

(iii) $A \cup (B \cup C) = (A \cup B) \cup C$,

$A \cap (B \cap C) = (A \cap B) \cap C$.

(iv) $(A \cup B)' = A' \cap B'$,

$(A \cap B)' = A' \cup B'$.

(v) $A \cup (B \cap C) = (A \cup B) \cap (A \cup C)$,

$A \cap (B \cup C) = (A \cap B) \cup (A \cap C)$.

For example, here is how (iv) is proved:

$$x \in (A \cup B)' \equiv \sim (x \in A \cup B)$$
$$\equiv \sim ((x \in A) \vee (x \in B))$$
$$\equiv (\sim x \in A) \,\&\, (\sim x \in B) \quad \text{(tautology)}$$
$$\equiv x \in A' \,\&\, x \in B'$$
$$\equiv x \in A' \cap B'.$$

The remaining identities are proved in a similar way.

Frequently we wish to operate not with two sets, but with an arbitrary, perhaps infinite, family of sets. Let $\{A_\lambda \mid \lambda \in L\}$ denote an arbitrary family of subsets of X; the set L is called the *"index set."* Then we can define:

1. *Union:* $\bigcup_{\lambda \in L} A_\lambda = \{x \in X \mid \exists \, \lambda \in L, x \in A_\lambda\}$.

2. *Intersection:* $\bigcap_{\lambda \in L} A_\lambda = \{x \in X \mid \forall \, \lambda \in L, x \in A_\lambda\}$.

In other words, x belongs to the union of the family of sets A_λ if and only if x belongs to at least one set A_λ; and x belongs to the intersection of the family of sets A_λ if and only if x belongs to every set A_λ.

Using the quantifier calculus of Section I.2, we can obtain analogues of the identities (iii), (iv), and (v) for the case of arbitrary families of sets; see Exercise 3.

Finally let us give the modern set-theoretic definition of "function."

Definition 2   Let A and B be given (nonempty) sets. A set f of ordered pairs (a, b) with $a \in A$ and $b \in B$, is called a **function** from A to B, provided that for every $a \in A$ there exists a unique $b \in B$ such that $(a, b) \in f$. In case $(a, b) \in f$, we write $f(a) = b$.

The set A is called the *domain* of f, and B is called the *codomain* of f. If f is a function from A to B, we write

$$f : A \to B.$$

Example

Let $A = B = \mathbb{R}$ and consider the set f of all ordered pairs (x, x^2) for $x \in \mathbb{R}$. This set satisfies the definition of a function, and we write $f(x) = x^2$.

Definition 2, which may seem mysterious on first sight, actually agrees with most people's understanding of "function." (Most calculus books, however, refer to the set of pairs $(a, b) \in f$ as the *graph* of the function.)

Exercises

1. Prove the set identities (i), (ii), (iii), and (v), identifying the tautologies used.

2. Let $A_\lambda = \left\{ x \in \mathbb{R} \mid \dfrac{1}{\lambda} \leq x \leq \lambda \right\}$. If

$$L_1 = \{\lambda \in \mathbb{R} \mid \lambda > 0\} \quad \text{and} \quad L_2 = \{\lambda \in \mathbb{R} \mid 2 < \lambda < 5\},$$

find

$$\bigcap_{\lambda \in L_i} A_\lambda \quad \text{and} \quad \bigcup_{\lambda \in L_i} A_\lambda \quad (i = 1, 2).$$

3. Prove the following statements.

(a) $\left(\displaystyle\bigcup_{\lambda \in L} A_\lambda \right)' = \bigcap_{\lambda \in L} A_\lambda'$,

$\left(\displaystyle\bigcap_{\lambda \in L} A_\lambda \right)' = \bigcup_{\lambda \in L} A_\lambda'$.

(b) $\left(\displaystyle\bigcup_{\lambda \in L} A_\lambda \right) \cap B = \bigcup_{\lambda \in L} (A_\lambda \cap B)$.

4. $\displaystyle\bigcup_{\lambda \in \phi} A_\lambda = ?$ $\displaystyle\bigcap_{\lambda \in \phi} A_\lambda = ?$

5. Let f be the set of all pairs (x, y) with $x \in \mathbb{R}$, and with

$$y = \begin{cases} 1 & \text{if the digit 7 appears in the decimal} \\ & \text{expansion of } x, \\ 0 & \text{otherwise.} \end{cases}$$

Calculate $f(\frac{1}{9}), f(\frac{3}{7}), f(\sqrt{2})$.

6. Let $\mathscr{S}$ denote the set of all sets. Consider the set

$$T = \{x \in \mathscr{S} \mid \sim(x \in x)\}.$$

Show that $T \in T \Leftrightarrow \sim(T \in T)$.

This is called *Russell's paradox*, after Bertrand Russell (1872–1969). The moral is, roughly speaking, that the "set" of all sets isn't really a set itself because it's too "inconceivable." Set theory is today a major area of mathematical research, and there have been recent revolutionary developments.

II Mathematical Induction

II.1 Mathematical Induction

Consider the following simple problem.

Example 1

Find the value of

$$1 + 4 + 9 + 16 + \cdots + 100.$$

(What are the missing terms, represented by 3 dots?) Now find the value of

$$1 + 4 + 9 + 16 + \cdots + 10{,}000.$$

No sane person would try to solve this by arithmetic—he'd look for a more "mathematical" method. If he looks in the right book (for example, this one) he'll find the following formula:

$$1^2 + 2^2 + 3^2 + \cdots + n^2 = \tfrac{1}{6}n(n+1)(2n+1). \qquad \text{(II.1)}$$

The values of the sums given above are 385 and 338,350, respectively.

The question now is, how can Formula (II.1) be derived? The truth is that it's tricky to *discover* such formulas, although guessing sometimes works. Let's suppose we obtain (II.1) by guessing. Next question: is the guess correct?

First we can check a few special cases (we may soon discover if (II.1) is *wrong!*).

n	1	2	3	5	10	etc.
$1^2 + 2^2 + 3^2 + \cdots + n^2$	1	5	14	55	385	?
$\frac{1}{6}n(n+1)(2n+1)$	1	5	14	55	385	?

Is this table a *proof* of Formula (II.1)? *No.* (Remember the "physicists' proof" that every odd integer is a prime: "3 is a prime, 5 is a prime, 7 is also, 9 is an experimental error, but 11 is prime, 13 is prime, and so on.")

Here's how to prove (II.1); note that we are trying to prove that (II.1) *holds for every positive integer n.* Suppose not. Then there exists a positive integer N_0 such that (II.1) is false for $n = N_0$. Let N denote the first (smallest) positive integer for which (II.1) is false. Then:

(a) $N > 1$, because (II.1) is surely true if $N = 1$.
(b) Formula (II.1) is true for $n = N - 1$.
(c) It's false for $n = N$.

Accordingly, by (b).

$$1^2 + 2^2 + 3^2 + \cdots + (N-1)^2 = \tfrac{1}{6}(N-1)N(2(N-1)+1).$$

Adding N^2 to both sides, we have

$$1^2 + 2^2 + 3^2 + \cdots + (N-1)^2 + N^2 = \tfrac{1}{6}(N-1)N(2N-1) + N^2$$
$$= \tfrac{1}{6}N[(N-1)(2N-1) + 6N]$$
$$= \tfrac{1}{6}N[2N^2 + 3N + 1]$$
$$= \tfrac{1}{6}N(N+1)(2N+1).$$

But this shows that (II.1) is true for $n = N$, in contradiction to (c). Therefore, we have proved that Formula (II.1) must hold for every n.

The proof just given is an *indirect proof by mathematical induction.* ("Indirect" means by contradiction.) We could have given a direct proof, as in Example 3 below.

Example 2

It is well known that the sum of the interior angles of any triangle is 180°—see Figure A.1.

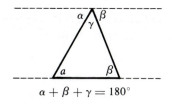

$$\alpha + \beta + \gamma = 180°$$

Figure A.1

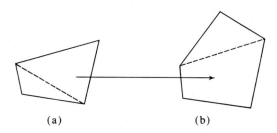

(a)   (b)

Figure A.2

Question: What is the sum of the interior angles of a convex polygon with n sides? A good student should be able to figure this out, so read no further until you've tried.

Now let us try to guess the answer. Take $n = 4$ first. By Figure A.2(a), the sum of the interior angles is obviously $2 \times 180°$ (why?). But now take $n = 5$ (Figure A.2(b)). The interior angles obviously total $3 \times 180°$. We can guess that generally:

> *The sum of the interior angles of an n-sided convex polygon* ($n \geq 3$) *equals* $(n - 2) \times 180°$.

But this is more than a guess—we can easily see why it is true for every n. Namely, every time we increase n to $n + 1$ we add another triangle, so that the interior angles increase by $180°$. Since the formula $(n - 2) \times 180°$

(a) is correct for $n = 3$, and
(b) is correct for any value $n + 1$ whenever it is true for n,

we must conclude that this formula is correct for *all* values of n, $n \geq 3$.

The foregoing argument is a simple example of a (direct) proof by mathematical induction. In general, the method of proof by mathematical induction is based on the following axiom.

Axiom of Mathematical Induction   *Let P_n represent a formula, or other statement, concerning the positive integer n. Then P_n is true for all positive integers n provided that*

   (i)   *P_1 is true, and*
   (ii)   *whenever P_k is true (for some $k \geq 1$), then so is P_{k+1}.†*

We call this an axiom because it cannot be proved on the basis of simpler axioms. (For an indirect formulation of the axiom of mathematical induction, see Exercises 15 and 16.) Most students will probably agree, at least after a little study, that the axiom is intuitively "obvious"—it really describes a property of the system $\{1, 2, 3, \ldots\}$ of positive integers which we automatically believe. Here is how the axiom of mathematical induction is used in practice.

Example 3

$$1 + 2 + 3 + \cdots + n = \tfrac{1}{2}n(n + 1). \tag{II.2}$$

Let us prove this famous formula by induction:‡

   (i) P_1 is true, because P_1 says $1 = \tfrac{1}{2} \cdot 1 \cdot (2)$.
   (ii)   if P_k is true, then

$$1 + 2 + 3 + \cdots + k = \tfrac{1}{2}k(k + 1).$$

Adding $k + 1$ to both sides, we have

$$1 + 2 + 3 + \cdots + k + (k + 1) = \tfrac{1}{2}k(k + 1) + (k + 1)$$
$$= \tfrac{1}{2}(k + 1)(k + 2).$$

This proves that P_{k+1} is true (why?). Therefore by the axiom, P_n is true for all n. Notice how similar this direct proof is to the indirect proof we gave in Example 1.

Let us repeat: in order to prove the statement "P_n is true for all integers $n \geq 1$" (that is, "$\forall\, n \geq 1,\ P_n$") by mathematical induction, we must do the following:

   (i)   check that P_1 is true,
   (ii)   prove that for every integer $k \geq 1$, if P_k is true then so is P_{k+1}.

---

† In precise logical terms, Condition (ii) says: for all $k \geq 1$, P_k implies P_{k+1}.
‡ "Induction" means the same as "mathematical induction."

In the exercises below, we use the *summation notation*. The expression $\sum_{k=n}^{m} a_k$ denotes the sum of a_k from $k = n$ to $k = m$;

$$\sum_{k=n}^{m} a_k = a_n + a_{n+1} + \cdots + a_m. \tag{II.3}$$

Exercises

1. Write out in full (every term of) the following sums.

    (a) $\displaystyle\sum_{j=1}^{4} \frac{1}{j}$,
    (b) $\displaystyle\sum_{k=5}^{8} 2^k$,

    (c) $\displaystyle\sum_{p=3}^{3} \sin \frac{p\pi}{2}$,
    (d) $\displaystyle\sum_{m=0}^{10} 1$.

2. Evaluate each sum in Exercise 1.

3. Show that

$$\sum_{k=m}^{n} a_k = \sum_{j=m}^{n} a_j.$$

Therefore, the variable of summation is a "dummy" variable. It's like x in $\int_a^b f(x)\, dx$.

4. Write out an "indirect" proof of Formula (II.2).

5. Note that
$$1 = 1,$$
$$1 - 4 = -(1 + 2),$$
$$1 - 4 + 9 = 1 + 2 + 3,$$
$$1 - 4 + 9 - 16 = -(1 + 2 + 3 + 4).$$

Guess the general law, and prove it by induction.

6. The following summation formulas are to be proved by induction. You may wish to write out some of the sums longhand—for example,

$$\sum_{k=1}^{n} (2k - 1) = 1 + 3 + 5 + \cdots + (2n - 1).$$

    (a) $\displaystyle\sum_{k=1}^{n} (2k - 1) = n^2$.
    (b) $\displaystyle\sum_{k=1}^{n} k^3 = \left\{ \sum_{k=1}^{n} k \right\}^2$.

    (c) $\displaystyle\sum_{k=0}^{n} a^k = \frac{1 - a^{n+1}}{1 - a}$ $(a \neq 1)$.

7. Show that for $a \neq 1$

$$\sum_{k=1}^{n} ka^k = \frac{a}{(1-a)^2}[na^{n+1} - (n+1)a^n + 1].$$

(Can you do this in two ways, one not using induction?)

8. Prove that if $0 < x < 1$, then $0 < x^n < 1$ for every positive integer n.

9. Prove that if $\varepsilon > 0$, then

$$(1+\varepsilon)^n \geq 1 + n\varepsilon \quad (n = 1, 2, 3, \ldots).$$

10. Show that if n straight lines are drawn in the plane, then the total number of intersections cannot exceed $\frac{1}{2}n(n-1)$.

11. Using the product rule of calculus, $D(uv) = uD(v) + vD(u)$, show by induction that

$$D(u^n) = nu^{n-1}D(u) \quad (n = 1, 2, 3, \ldots).$$

12. For which integer n is $2^n > n^2$? Prove your answer by induction.

13. Show that every positive integer is interesting.

14. Inductive proofs don't always have to begin with $n = 1$; often we wish to prove a statement: P_n for all $n \geq N_0$. (See Example 2 and Exercise 12.) State a more general form of the axiom of mathematical induction to cover such cases. Can you prove that the general form is a consequence of the original form?

15. Consider the following axiom:

    Axiom X   *Every nonempty subset S of the set of all positive integers contains a least element.*

    Prove that Axiom X implies the axiom of mathematical induction. (*Hint:* Let P_n be a given statement and suppose P_n satisfies conditions (i) and (ii) of the axiom of mathematical induction. Suppose that P_n is not true for all integers $n \geq 1$. Let S denote the set of integers for which P_n is false. Obtain a contradiction.) Note that Axiom X was used implicitly in Example 1 of this section.

*16. Show that the axiom of mathematical induction implies Axiom X, so that these two axioms are logically equivalent.

*17. Determine, as a function of n, the maximum possible number of subregions of the plane which can be formed by drawing n straight lines (extending to infinity in both directions) in the plane. (It's not 2^n.)

*18. Show that the sum of the digits in any multiple of 9 is divisible by 9.

19. n straight lines (extending to infinity in both directions) are drawn in the plane. The resulting configuration is to be colored like a map—no two "countries" with a common border may have the same color (but two countries which meet at a single point can have the same color). Prove that only two colors are needed.

II.2 Mathematical Induction (Continued)

It is often convenient to use mathematical induction in definitions.

Example

Let a sequence $\{x_n\}$ be defined by the following "recursion formula" (see Chapter 1).

$$x_1 = 1, \qquad x_{n+1} = 2x_n + 1 \quad (n \geq 1).$$

It is easy to calculate a few values of x_n and then to guess and prove the general formula:

$$x_1 = 1, \qquad x_2 = 3, \qquad x_3 = 7, \qquad x_4 = 15,$$

and we can guess that in general $x_n = 2^n - 1$. We leave it to the reader to prove this guess by induction (or to disprove it).

A sequence defined by means of a recursion formula is sometimes said to be "defined inductively." Some further examples are given in the exercises.

Next we consider another form of the axiom of mathematical induction which is sometimes useful.

Axiom of Mathematical Induction—Strong Form   *A statement P_n is true for all integers n provided that*

(i) *P_1 is true, and*

(ii) *whenever P_j is true for all $j < k$, then so is P_k true (for $k = 2, 3, 4, \ldots$).*

This is called the "strong form," because it is generally easier to verify (ii) in this form than (ii) of the earlier axiom. Here is an example; this theorem would be very hard to prove using the first axiom directly.

Theorem   *Every integer ≥ 2 can be written as a product of primes.*

Proof   (Remember that, by definition, n is a *prime* if and only if the only integers which divide n are 1 and n.) Consider the statement P_n: the integer n is a product of primes. Obviously P_2 is true. To prove that (ii) of the strong form of the axiom of mathematical induction holds, let k be a given integer > 2 and suppose that every integer $j < k$ is a product of primes. There are two cases:

(a) if k itself is a prime, then P_k is true.

(b) if k is not a prime, then $k = l \cdot m$, where l and m are $< k$. Hence, by assumption, both l and m are products of primes. Therefore so is k a product of primes.

This proves (ii) and therefore we conclude that P_n is true for all n. ∎

Exercises

1. Define x_n $(n = 1, 2, 3, \ldots)$ by

$$x_1 = 1; \qquad x_{n+1} = x_n - \frac{1}{n(n+1)}.$$

Determine x_n explicitly.

2. Define $\{x_n\}$ as follows:

$$x_1 = 1; \qquad x_{n+1} = \frac{1}{1 + x_n} \quad (n \geq 1).$$

Check that the first few terms of this sequence are $\{1, \frac{1}{2}, \frac{2}{3}, \frac{3}{5}, \frac{5}{8}, \ldots\}$. Now show by induction that

$$x_n = \frac{Z_n}{Z_{n+1}},$$

where $\{Z_n\}$ is the Fibonacci sequence, which is defined inductively as follows:

$$Z_1 = Z_2 = 1; \qquad Z_{n+2} = Z_{n+1} + Z_n \quad (n \geq 1).$$

3. Define the symbol $\sum_{k=1}^{n} a_k$ by induction. (The sequence $\{a_k\}$ is assumed given in advance.)

4. Define (x_n) by

$$x_1 = 1: \qquad\qquad x_{n+1} = \frac{x_n^2 + 2}{2x_n}$$

Show that every x_n is a rational number.

5. Define $x_1 = 2$ and $x_{n+1} = 2^{x_n}$ $(n \geq 1)$. Find x_2, x_3, x_4. Show that x_6 cannot be calculated by hand in a lifetime.

6. Prove the strong form of the axiom of mathematical induction from the original form (Section II.1). (*Hint:* Let P_n satisfy the hypotheses of the strong form, and let Q_n denote the statement "P_k is true for $1 \leq k \leq n$." Show that Q_n is true for all n.)

II.3   The Binomial Theorem

Wanted: a formula for $(a + b)^n$, valid for all positive integers n. Examples:

$$(a + b)^1 = a + b,$$
$$(a + b)^2 = a^2 + 2ab + b^2,$$
$$(a + b)^3 = a^3 + 3a^2b + 3ab^2 + b^3,$$
$$(a + b)^4 = a^4 + 4a^3b + 6a^2b^2 + 4ab^3 + b^4.$$

The general formula will obviously be of the form

$$(a + b)^n = \sum_{k=0}^{n} C_k a^{n-k} b^k;$$

the problem is to determine C_k.

Notation: (1) If n is a positive integer, $n!$ denotes

$$n! = 1 \cdot 2 \cdot 3 \cdots n;$$

also by definition

$$0! = 1.$$

(This strange definition is to make the binomial theorem simple.)
   (2) If $0 \leq k \leq n$, both positive integers, then

$$\binom{n}{k} = \frac{n!}{k! \, (n - k)!}.$$

These numbers are called "binomial coefficients" because (we will show) $C_k = \binom{n}{k}$.

Lemma

$$\binom{n}{k} + \binom{n}{k-1} = \binom{n+1}{k}. \tag{II.4}$$

The proof (simple algebra) is left to you—see Exercise 2.

The Binomial Theorem

$$(a+b)^n = \sum_{k=0}^{n} \binom{n}{k} a^{n-k} b^k. \tag{II.5}$$

In this formula it is assumed that a and b are real numbers, and n is a positive integer. The reader should check the formula for a few small values of n, such as $n = 3$ or 4.

Proof   By inspection, the formula is valid for the case $n = 1$. To complete the proof we use mathematical induction. Thus (changing notation slightly from Section II.1) let us suppose (II.5) holds for n. We must deduce that the formula holds for the next case, $n + 1$. We have

$$(a+b)^{n+1} = (a+b)(a+b)^n$$

$$= (a+b)\sum_{k=0}^{n} \binom{n}{k} a^{n-k} b^k \quad \text{(by assumption)}$$

$$= \sum_{k=0}^{n} \binom{n}{k} a^{n+1-k} b^k + \sum_{k=0}^{n} \binom{n}{k} a^{n-k} b^{k+1}$$

$$= \text{first sum} + \sum_{k=1}^{n+1} \binom{n}{k-1} a^{n+1-k} b^k \quad \text{(why?)}$$

$$= \binom{n}{0} a^{n+1} + \sum_{k=1}^{n} \left[\binom{n}{k} + \binom{n}{k-1} \right] a^{n+1-k} b^k + \binom{n}{n} b^{n+1}$$

$$= a^{n+1} + \sum_{k=1}^{n} \binom{n+1}{k} a^{n+1-k} b^k + b^{n+1} \quad \text{(by lemma)}$$

$$= \sum_{k=0}^{n+1} \binom{n+1}{k} a^{n+1-k} b^k.$$

This is indeed the formula (II.5) for the case $n + 1$, and the proof is complete. ∎

Example

Find the coefficient of x^5 in the expansion of $(2 - x)^8$.

Solution   The term in question is

$$\binom{8}{5} 2^3(-x)^5 = \frac{8!}{5!\, 3!}\, 8(-1)^5 x^5$$

$$= -448x^5;$$

the coefficient is -448.

There are other proofs of the binomial theorem. It is a simple consequence of Taylor's theorem from the calculus, for example. The coefficients $\binom{n}{k}$ are sometimes denoted by $_nC_k$, and $_nC_k$ is equal to the number of combinations of n objects taken k at a time. A proof of the binomial theorem may be built on the study of permutations and combinations.

Pascal's triangle (below) has a simple relationship with the binomial theorem, which we leave the reader to elucidate. There is a generalization of the binomial theorem that holds if n is not an integer. The sum in (II.5) becomes an infinite series in general (see Section 5.9, Exercise 8).

					1						$(n = 0)$
				1		1					$(n = 1)$
			1		2		1				$(n = 2)$
Pascal's triangle		1		3		3		1			$(n = 3)$
	1		4		6		4		1		$(n = 4)$
1		5		10		10		5		1	$(n = 5)$
					$\cdots$						$\cdots$

Exercises

1. Verify the following formulas.

(a) $\binom{n}{0} = \binom{n}{n} = 1.$

(b) $\binom{n}{1} = \binom{n}{n-1} = n.$

(c) $\dfrac{n!}{k!} = n(n-1)\cdots(k+1)$ if $k \le n.$

(d) $\dfrac{k!}{k} = (k-1)!.$

(e) $\binom{n}{k} = \binom{n}{n-k}.$

2. Prove the Formula (II.4).

3. Find

    (a) the 4th term in the expansion of $(\alpha + \varepsilon\beta)^{10}$,

    (b) the constant term in the expansion of $\left(x^2 - \dfrac{1}{x^2}\right)^8$,

    (c) the coefficient of x^4 in the expansion of $(x^2 + y)^7$.

4. Estimate $(1.02)^5$ to 4 decimals.

5. Prove that

$$\binom{n}{0} + \binom{n}{1} + \binom{n}{2} + \cdots + \binom{n}{n} = 2^n.$$

*6. Prove the following formula, called *Leibniz's rule.*

$$D^n(uv) = \sum_{k=0}^{n} \binom{n}{k} D^k u\, D^{n-k} v.$$

$$\left(D^n \text{ stands for the } n\text{th derivative, that is, } D^n w = \frac{d^n w}{dx^n}\, .\right)$$

*7. State and prove the trinomial theorem. (*Hint:* It is helpful to rewrite the binomial theorem in the form

$$(a + b)^n = \sum_{j+k=n} \frac{n!}{j!\,k!} a^j b^k$$

where $j \geq 0, k \geq 0$.)

Solutions to Selected Exercises

Solutions are given only for exercises that have explicit numerical or logical answers, and not for exercises requiring proofs or estimates.

Chapter 1

Section 1.1

1. (a) $2, 4, 8, 16, \ldots$,   (b) $-1, +1, -1, \ldots$,   (c) $0, 1, 0, 1, \ldots$,
   (d) $-1, -2, -3, -2, -1, 0, +1, \ldots$.

2. (d) $x_n = n^2 + 1$,   (e) $x_n = 2^n + 2$,   (f) $x_n = n!$ (n factorial).

3. (a) $x_n = 2n - 1$,   (b) $x_n = \frac{1}{2}\{(-1)^{n+1} + 1\}$,   (c) $x_n = 2^{n-2} (n \geq 2)$,
   (d) $x_n = n$,   (e) $x_n = (-1)^{n+1}$.

Section 1.2

1. $x_{n+1} = (x_n^2 + a)/(2x_n)$.

4. If $x_1 = 1$, the sequence $\{x_n\}$ is $\{1, 2, 1, 2, \ldots\}$, which diverges.

6. $x_{n+1} = (2x_n^3 + 2)/(3x_n^2)$.

Section 1.4

1. (a) 1,   (b) 1/4,   (c) diverges,   (d) diverges,   (e) 0,   (f) 0.

2. (a) -1,   (b) 1,   (c) 0,   (d) 0,   (e) 0.

Section 1.7

1. (a) 2, 4, 6,   (b) 2, 3, 4, 6,   (c) 4, 6,   (d) 1, 5, 7,   (e) 2, 4, 6.

10. No. Example: $x_n = 0, y_n = \dfrac{1}{n}$.

Section 1.10

4. $+\infty$.

Chapter 2

Section 2.2

1. (a), (k), (l), and (n) diverge; the rest converge.

Section 2.3

1. (a) diverges; (b) converges conditionally; (c) converges absolutely; (d) converges conditionally; (e) converges absolutely; (f) converges absolutely; (g) converges absolutely; (h) diverges.

Section 2.4

1. (a) 1; (b) 1/2; (c) $+\infty$; (d) 1/e.

3. (a) diverges at $x = \pm 1$; (b) diverges at $x = +1$, converges conditionally at $x = -1$; (c) converges absolutely at $x = \pm 1$.

Chapter 3

Section 3.1

3. (a) 2/3,   (b) 2,   (c) 0,   (d) 0.

Section 3.3

1. (a) $-\infty$,   (b) $+\infty$,   (c) $-1/2$,   (d) $+\infty$,   (e) 0,   (f) 0,

(g) $\pi/2$,   (h) $-\infty$.

2. (a) 1,   (b) 0,   (c) 1/2,   (d) 1/2.

Section 3.4

1. (a) Infinite discontinuities at $x = \pm 1$.   (b) Infinite discontinuity at $x = -1$; removable discontinuity at $x = +1$, with limit 1.   (c) Jump discontinuity at $x = 1$, with $J_1(f) = 2$.   (d) Infinite discontinuities at $x = n\pi$, $n = 0, \pm 1, \pm 2, \ldots$.   (e) Removable discontinuity at $x = 0$, with limit 0.   (f) Same as (e).   (g) Jump discontinuity (jump 1) at every point $x = n/10$, $n = 0, \pm 1, \pm 2, \ldots$.   (h) Infinite discontinuity at $x = 0$.   (i) Removable discontinuity at $x = 0$, with limit 0; infinite discontinuities at $x = \pm 1$.   (j) Infinite oscillating discontinuity at $x = 0$.

3. Removable.

Section 3.5

2. (a) $x = 0$,   (b) $x = 0$,   (c) $x = n\pi$, $n = 0, \pm 1, \pm 2, \ldots$,

(d) $x = 0$ and $x = 2$.

4. (a) $-1, +1$,   (b) $+1, -1$,   (c) $+1, -1$,

(d) $0, -2$ at $x = 0$, $+2, 0$ at $x = 2$.

Chapter 4

Section 4.2

1. (a) $\inf A = \min A = 0$, $\sup A = +\infty$, max A doesn't exist,

    (b) $\inf A = \min A = 7\pi/22$, $\sup A = \max A = 10$,

    (c) $\inf A = 0$, $\sup A = 1$, neither min A nor max A exists,

    (d) $\inf A = 0$, $\sup A = \max A = 1$,

    (e) $\inf A = 0$, $\sup A = \max A = 0.888 \cdots = 8/9$.

3. (a) 1 (not max),   (b) 0 (not min),   (c) $2 = \min$,

    (d) $+1$ (not max), -1 (not min),   (e) 1 (not max),   (f) $0 = \min$,

    (g) min $= \sin x_0/x_0$, where x_0 is the first positive solution of the transcendental equation $\tan x = x$ $(x_0 \approx 4.493$, min $\approx -0.217)$.

6. $\inf A = \min A = 1$, $\sup A = \max A = 3^{1/3} = 1.442 \cdots$.

7. (a) 3 (in both cases).

Section 4.3

1. (a) $[-1, +1]$,   (b) The empty set,   (c) $\mathbb{R}$,   (d) $\{0\}$.

Chapter 5

Section 5.1

1. (a) Relative maxima at $x = 0$ and $x = 1$; relative minimum at $x = 2/3$.

    (b) Relative maxima at $x = 0$ and $x = 2$; relative minima at $x = -1$ and $x = 2/3$.

    (c) Relative maximum at $x = 1$; relative minima at $x = 0$ and $x = 2$.

    (d) Relative maxima at $x = 1/2$ and $x = 2$; relative minima at $x = 0$ and $x = 1$.

3. $\xi = a/4$; yes.

Section 5.2

1. $U_\pi(f) = (n + 1)(2n + 1)/6n^2$; $L_\pi(f) = (n - 1)(2n - 1)/6n^2$;
    limits are both 1/3.

2. Let $\delta = \min(1/2, \varepsilon/11)$ and take $\pi = \{0, 1 - \delta, 1, 2 - \delta, 2, \ldots, 10 - \delta, 10\}$.

3. $\sum_{k=1}^{n} c_k \Delta x_k.$

Section 5.3

1. (a) $2e^{-4x^2}$,   (b) $2e^{-4x^2} - e^{-x^2}$.

3. $G(x) = x$ if $0 \le x \le 1$,   and $G(x) = 2x - 1$ if $1 \le x \le 2$. Notice that $G'(1)$ does not exist.

Section 5.4

1. $p > -1$.

2. $p < -1$.

5. (a) Diverges,   (b) Converges,   (c) Converges,   (d) Converges,
    (e) Converges.

6. (a) π,   (b) 1/2.

Section 5.5

1. (a) $f_n \to 0$ (pointwise); the convergence is uniform on I_2 only,

    (b) $f_n(x) \to 1$ if $x \ne 0$, and $f_n(0) \to 0$; the convergence is not uniform on any of the intervals,

    (c) $f_n(x) \to 0$ if $x \ne 0$, and $f_n(0) \to 1$; the convergence is uniform on I_3 only.

2. (a) $f(x) \equiv 0$,   (b) $[n/(n + 1)]^{n+1}$,   (c) No,   (d) Yes.

Section 5.6

1. (a) 2, (b) 1, (c) $+\infty$, (d) 2, (e) 2/3.

2. (a) 1; (b) 1/2; (c) 1/2: (d) 1. The limit (5.24) exists for (a) and (b) only.

Section 5.7

1. (a) $\sum_{k=0}^{\infty} (-1)^k x^{2k+1} \Big/ (2k+1)!$ ($|x| < 1$).

(b) $\sum_{k=0}^{\infty} (-1)^k x^{2k} \Big/ (2k)!$ ($|x| < 1$).

(c) $\sum_{k=0}^{\infty} x^{2k+1}/(2k+1)!.$

3. $\sum_{0}^{\infty} a_j x^j$, where $a_{2k} = (-1)^k (\sin x_0)/(2k)!$ and $a_{2k+1} = (-1)^k (\cos x_0)/(2k+1)!$

4. (a) $x/(1-x)^2$,   (b) $x(x+1)/(1-x)^3$.

7. (b) 0.90484.

8. 0.747

Chapter 6

Section 6.1

1. $\|x\| = \sqrt{10}$; $\|y\| = \sqrt{6}$; $\|x + y\| = 2\sqrt{5}$; $\sum x^i y^i = 2$.

4. (a) Closed,   (b) Closed,   (c) Open,   (d) Neither,
(e) Neither,   (f) Closed.

Section 6.2

5. (a) $\{x \in \mathbb{R}^3 \mid x_1 = 0\}$,   (b) $\varnothing$,   (c) $\mathbb{R}^n$,   (d) $\{0\}$,   (e) $\mathbb{R}^2$,
(f) $\bar{N}_r(0) \cap \bar{H}$.

Section 6.3

3. (a) Two turns of a conical helix,   (b) An ellipse,   (c) A straight line.

4. [This is a *torus*.]

5. $f(t) = (1 - t)x_1 + tx_2, \ 0 \le t \le 1.$

Section 6.4

1. (a) $f_{max} = f(1, -1) = 4; \ f_{min} = f(-1, 1) = -4,$

2. (b) $f_{max} = f(1/\sqrt{2}, 1/\sqrt{2}) = f(-1/\sqrt{2}, -1/\sqrt{2}) = 1/2;$
    $f_{min} = f(1/\sqrt{2}, +1/\sqrt{2}) = f(-1/\sqrt{2}, 1/\sqrt{2}) = -1/2.$

    (c) $f_{min} = f(x, 0) = f(0, y) = 0$ (for arbitrary $x, y \ge 0$); $f_{max} = f(1/\sqrt{2}, 1/\sqrt{2}) = 1/2e,$

    (d) $f_{min} = f(1/5, 2/5) = 1/5;$ there is no max.

Appendix I

Section 1

6. The *last columns* in the truth tables are:
    (a) FTTF,   (b) TTTF (same as $\sim B \Rightarrow A$),
    (c) TFTT (same as $B \Rightarrow A$),   (d) TTFT (same as $A \Rightarrow B$),
    (e) This is not a logical assertion.

Section 2

1. (a) T,   (b) T,   (c) F,   (d) T,   (e) T,   (f) T.

3. (a) F if S is empty, otherwise T,   (b) T,   (c) F,   (d) F,
    (e) T,   (f) T,   (g) T.

Section 4

2. (a) $\underset{L_1}{\bigcup} A_\lambda = (0, \infty)$,   (b) $\underset{L_1}{\bigcap} A_\lambda = \emptyset$,   (c) $\underset{L_2}{\bigcup} A_\lambda = (\frac{1}{5}, 5)$,

    (d) $\underset{L_2}{\bigcap} A_\lambda = [\frac{1}{2}, 2]$.

5. (a) $f(\frac{1}{9}) = 0$,   (b) $f(\frac{3}{7}) = 1$,

    (c) $f(\sqrt{2}) = 1$, since $\sqrt{2} = 1.414213562373\ldots$.

Index